Towards a Social Ecology
F. E. Emery and E. L. Trist

The authors of this book, Fred E. Emery and Eric L. Trist, are social psychologists who have become involved in wider organizational and environmental studies. The interest which first brought them together was the development of a socio-technical approach to the study of organizations considered as open systems in relation to their environments. In collaborative studies of systems larger than that of the single organization, they found they had to extend the open system concept to include the idea of the social environment as a quasi-independent domain. This led them to introduce the notion of *The Causal Texture of Organizational Environments*, a joint paper published in *Human Relations* in 1965. Concerned with the implication for human adaptive capability of the increasing 'turbulence' of the world environment caused by the accelerating rate of change and the associated rises in the levels of complexity, inter-dependence and uncertainty, the authors have fully developed this line of thought in *Towards a Social Ecology*.

Part I by Emery is a full-length exposition of ideas briefly outlined in a paper prepared for the Committee on the Next Thirty Years of the British Social Science Research Council. Part II by Trist uses these ideas in analysing several critical aspects of the transition to the post-industrial society. Concepts and methodologies for improving our capability actively to choose our future rather than passively to accept whatever may turn up are discussed in Part I and applied in Part II with particular reference to changes in values.

:cology
:e Present

Towards A Social Ecology

Contextual Appreciation of the Future in the Present

by F E Emery and E L Trist

Plenum Press · London · 1972

Plenum Publishing Company Ltd.
Davis House
8 Scrubs Lane
Harlesden
London NW10 6SE
Tel. 01-969-4727

SBN: 306 30563-1
Library of Congress Catalog Card Number: 70-178778

Printed in Great Britain by
The Whitefriars Press Ltd., London and Tonbridge

Contents

Foreword by Sir Geoffrey Vickers vi
Editor's Preface x
Introduction xi

Part One
The next thirty years: concepts, methods and anticipations
by F E Emery

1 Planning and a conceptual basis for predicting the future 3
2 Complexity reduction 16
3 The early detection of emergent processes 24
4 The general characteristics of social fields: Environmental levels 38
5 Adaptation to turbulent environments 57
6 Active adaptation: The emergence of ideal seeking systems 68

Part Two
Aspects of the transition to post-industrialism
by E L Trist

7 Re-evaluating the role of science 83
8 The establishment of problem-oriented research domains 91
9 Collaborative social and technical innovation 103
10 The relation of welfare and development: systems and ideo-existential aspects 120
11 The relation of welfare and development: historical and contemporary aspects 137
12 The structural presence of the post-industrial society 156
13 The cultural absence of the post-industrial society 172
14 Task and contextual environments for new personal values 182
15 The risk-security balance and the burden of choice 190
16 From planning towards the surrender of power 203

Appendix–The socio-technical system as a source concept by F. E. Emery and E. L. Trist 211
References 224
Index 237

Foreword
Sir Geoffrey Vickers

This book is described as a contribution 'towards a social ecology'. As such it is timely and welcome. The phrase is not yet familiar, the concept still imprecise; we need urgently to move toward a better understanding of it.

The word 'ecology' began to become familiar outside scientific circles when human intervention in natural processes began to have effects so unforeseen, so dramatic and so disastrous as to make headline news. It might be infestation by an unfamiliar pest, like the rabbit in Australia; or soil erosion, from ill-controlled clearing and cropping; or pollution from fertilizers or industrial wastes; or urban proliferation; or toxic accumulation of pesticides. From the crescendo of such warnings, industrial man began to learn again what agricultural man learned long ago—that he is only one among many species, whose continued existence depends not only or primarily on competitive struggle but on most complex systems of mutual support, not less effective for being unconscious and unplanned. These are the kind of systems that ecologists study; so we look to ecology for light upon them.

The ecologist has a characteristic viewpoint. He looks at the total pattern of life in some defined habitat, in the belief that it constitutes one system. When he sees populations, of one species or another, growing or dwindling, oscillating or remaining strangely constant, he assumes that the regularities which make the pattern recognizable are due to the mutual influence which each population exercises, directly or indirectly, on all the others and all of them on their common physical environment. This net of relations is what he needs to understand; and the only assumption he can safely make is that it is a net—no mere tangle of causal chains but a field in which multiple, mutual influences are constantly at work. This is the interest and this the assumption which are slowly seeping through into the consciousness of Western man, despite the resistance of a culture drunk with the apparent success of exploitation accelerating over centuries.

The system which the ecologist studies is made up of many sub-systems, interdependent, overlapping or hierarchically organized. Each population, each group, each individual is, in varying degrees, organized as a system. As he attends first to one and then to another of these constituent sub-systems, he promotes it to the central role and turns the rest into 'environment'; but the choice is his and he can reverse it. He may have a special interest in one or another, a concern, for example, for breeding beef cattle or breeding grasses; but in either of those cases, he will be no less interested in the effect of grass on grazing than in the effect of grazing on grass. The human ecologist has declared his interest; man is his subject. But he too will often need to shift his focus and thus change the distribution of his field between figure and ground. Indeed, he needs to do so with especial subtlety. For he finds in his field a feature which other ecologists are largely spared. Whatever the human focus of his attention, be it individual, group, community, corporation or society, he finds that its environment is predominantly—human.

The relations of any human subject with its human surround are social or are frustrating because they are not social. Equally, the human subjects of his attention, individual or collective are social entities and would not be human if they were not. Any human ecology must be social ecology; but the name is especially apt to a study such as this, which is particularly concerned with human culture and human institutions, both as figure and as ground.

A culture, whether of a whole society or of some local or professional group, may be regarded as an aspect of the living people who embody it; but it may also be regarded as the soil in which they grow. It conditions them; yet they, to no small extent, can select, reject and change the heritage which they embody and pass on. Similarly, a society's institutions can be viewed either as part of its structure or as a framework which can constrain and imprison, as well as sustain; a human artifact, which men can remake. Both views are legitimate and both are needed.

An ecologist must choose his field by reference to his

interests and his problems. There are human-ecological problems which can only be seen with the human race as figure and the planet as ground. But most human-ecological problems call for a more restricted view. For all men's mobility and interaction, they still form relatively discrete societies. These divisions stem partly from the logistics of geography, not wholly abolished by aircraft and radio. One fifth of the world's water supply is generated in the basin of the Amazon; but this brings no relief to the ecological or the political problems associated with the waters of the Jordan. Political boundaries are equally divisive. Institutional means for distributing wealth across frontiers are radically different from those which distribute it within each political society, and far more limited.

The contrast should excite our surprise, not at what frontiers exclude, but at what they facilitate. The 'developed', crowded, urbanized state of today is an ecological miracle—or monstrosity—hard to credit and still harder to project into the future; a population of tens, even hundreds of millions, largely aggregated in cities, each individual dependent for his day to day existence on the activities of countless, largely unknown and indifferent strangers. It is possible only because institutions, often of immense size, have developed to structure this complexity, and only so long as culture makes these institutions work.

All these institutions, however narrowly defined their ostensible purpose, are themselves social organizations, a fact which the industrial age overlooked, to our cost even more than to its own. The authors of this book have played important parts in naturalizing the idea that a business organization is a *socio*-technical system, functioning as a whole only in so far as its social organization gives it inner coherence. The two requirements do not easily combine, for they involve conflicts of scale. Technological and economic factors promote growth to a size which social organization finds repugnant.

The same conflict besets the organization of those *socio-physical* systems we call cities. How to organize such a system on a scale large enough to deal with the activities that it generates—with its traffic flows, its flows of goods and services, its flow of information—and yet keep it, socially, to a human scale? The human ecologist is daily reminded that, through all but the last 2 per cent. of human history, men were accustomed to live in societies not larger than groups of families, and that this was probably true also for their pre-human ancestors through far greater gulfs of time. The whole agricultural epoch is contained within ten millenia, a mere 300 generations; and only from the middle of that epoch, in a few favoured areas, did wealth accumulate sufficiently to support the concentrations and to breed the problems of urban man.

Today, as the industrial age approaches what the authors regard as its close, these conflicts of scale have risen to a critical level. On the one hand, economic organization grows uncontrollably larger. On the other hand, an even fiercer demand arises for stable social organization on a scale small enough to be satisfying, meaningful and viable; a demand expressed by newly self-conscious groups, using with a new violence the new methods of coercion offered by societies whose stability is vulnerable as never before. These conflicts of scale are ever present to the social ecologist.

He is equally conscious of conflicts of scale in another dimension; for the ecologist, more than most, is conscious of the dimension of time. His concern is with events extended in time, and with processes of mutual influence which take time to accomplish. And wherever he fixes his attention on the human scene, he becomes aware of conflict between the fastest rate at which men and societies can be expected to change their responses to or even their understanding of their changing milieu, and the slowest rate at which that milieu can be expected to change. Human generations change no more quickly than before; indeed, with increased longevity, they may be said to change more slowly. But the changes which they unwittingly breed in their surround accelerate exponentially.

These papers should help to establish a way of thinking which is urgently needed and still far from general. Though the word 'system' is in constant use, its meaning is often restricted and its human implications still far from accepted. These include the acceptance of limitation; of mutual obligation; and of a sense of time which extends the present deep into the future as the concern of men now. And all this is implied as the inescapable consequence of the net of relations which alone can sustain our present societies, preserve our heritage and give our aspirations any hope of realization.

Introduction

By presenting jointly the essays brought together in this volume we hope that their context will be richer than if we had presented them separately. The ideas developed in Part One compose the conceptual theme on which the explorations of Part Two are based. That our efforts might lead to a book did not occur to us when the work was being done. This was suggested by Dr. J. D. Sutherland on one of the last occasions when the three of us met at the Tavistock Institute.

His suggestion brought home to the authors the incompleteness of what they had so far done. To make it into a book would entail a more thorough explication and development of the material than had been accomplished in the journal articles and working papers in which it existed. We decided to undertake this further task not only to follow through our own lines of thought but to make ourselves more intelligible to our fellow social scientists.

Clarification of our more recent work could remain only partial unless we re-examined what preceded it, namely the work on socio-technical systems and its relation to organization theory. It was the task of developing the socio-technical concept and carrying out associated empirical studies which first brought us together twenty years ago. But this field is now best reviewed from the perspective of the wider studies to which it has given rise. We have left it, therefore, to be the subject of a second volume, making this the first, in which, however, we have included an appendix which gives a brief description of the socio-technical system as a source concept.

As the decade of the 'fifties gave way to that of the 'sixties, projects being undertaken at the Tavistock Institute, in response to client needs, began to involve wider social systems than that of the single organization and to have a future orientation with a much longer time-horizon. Changes were taking place in the contextual as well as the task environment of organizations, and these were affecting the individual as a member of the social aggregate of his society in ways which the study neither of his personality nor of his immediate relations could fully account for. The levels of uncertainty and complexity were rising in consequence of the faster change rate brought on as the second industrial revolution, based on an information technology, gathered force. This lent greater importance to, as it created greater difficulties for, processes of planning and the consideration of alternative futures. We attempted to capture this new quality of the social field in a short joint paper presented to the International Congress of Psychology in Washington in 1963 entitled 'The Causal Texture of Organizational Environments'.

To take up such problems was to enter the field of *social ecology*. We had been led to it by our concern with what was happening to organizations, considered as open socio-technical systems, as they encountered greater complexity and a faster change-rate. Entailed was a more thorough examination than we had so far made of environmental relations and a consideration of the character of the environment itself. Problems of adapting to the more complex and rapidly changing environments which were becoming salient (we have called them 'turbulent fields') seemed to raise far-reaching questions concerning the future of man and society to which the social sciences should address themselves.

The stimulus to take the next step in our analysis came in the form of a request from Dr. Michael Young, then Chairman of the British Social Science Research Council, to prepare a preliminary report for the Committee on the Next Thirty Years on the conceptual and methodological problems of the social sciences in relation to forecasting. A review of what we had accomplished so far, and indeed of the literature, made it clear that a very considerable further theoretical development was needed. This was undertaken by Emery during the first six months of 1967, while Trist, who had gone to the University of California in Los Angeles, was concerned with developing the first university programme in socio-technical studies with Dr. Louis Davis who had been over at the Tavistock the previous year.

A first account of Emery's new thinking was published in

Human Relations in August 1967 and a second in *Forecasting and the Social Sciences* (1968), the volume which assembled the papers prepared for the Committee on the Next Thirty Years. These accounts have been revised and very substantially extended in the first six chapters of the present volume which will now be summarized.

Chapter 1 states the dilemma created by the awkward fact that conditions of rising complexity and uncertainty require the taking of an active role but make planning seem impossible. As trend projection is insufficient, a conceptual scheme, based on the dynamics of change is constructed by developing (a) Sommerhoff's theory of directive correlation and (b) a concept derived from Lewin of 'overlapping temporal gestalten'.

So great is the complexity of change that any estimation of future system states depends on finding methods of complexity-reduction (chapter 2). Given that the system can be identified and its characteristic generating function established, it may be possible to monitor changes in the system's 'values' and sometimes to identify the 'starting conditions of change'. But the greatest pay-off is likely to come from proceeding in terms of 'the leading part'.

A complementary methodology concerns the early detection of emergent processes (chapter 3). These lie concealed in existing systems with which they share parts. They may debilitate competing systems, intrude into them or bring about a state of 'mutual invasion'. As the growth process is sigmoid, too little time may be left for adaptation if an emergent process fails to be recognized until after it has surfaced. Changes in the state of symbolic systems can provide early warnings.

Western societies having been identified as 'the system' and the socio-technical organizations formed by the science-based industries as the leading part (chapter 4), a strategy of overall characterization of their environmental properties is proposed which yields four ideal types: the placid random, the placid clustered, the disturbed-reactive and the turbulent environments. Though these may all exist simultaneously, the last, in becoming salient, is posing new problems of adaptation.

The difficulties are so great that maladaptive defences are becoming massively in evidence. These represent different but related forms of splitting: superficiality in which depth connection is lost; segmentation in which parts pursue their ends without reference to the whole; and dissociation in which people and groups cease to respond to each other.

Nevertheless, adaptation to complex environments is possible by appropriate value transformations (chapter 5). Critical are the design principles on which social institutions are built. The choice is between the redundancy of parts (the machine principle) and the redundancy of functions (the organismic principle). The self-regulation and flexibility inherent in the latter give the possibility of adaptation to complexity and uncertainty.

Several favourable trends in the leading part may be currently observed (chapter 6). These include systems management, industrial democracy, matrix organizations, the institutional embodiment of wider social values, adaptive forms of planning, etc.

But there are also unfavourable trends arising from the maladaptive defences. These are producing conditions to which no adaptation is possible at all. They denote a fifth environment with characteristics of a vortex, signs of which already exist in certain parts of complex societies.

These ideas constitute the basis on which the studies reported in Part Two have been developed. They are called 'Aspects of the Transition to Post-Industrialism' as this designation identifies the process which is at the centre of contemporary social change.

Chapters 7-9 are concerned with changes in the leading part—the science-technology inter-face which has produced the second industrial revolution (the change-generating function). An anti-scientific ethos has appeared with regressive potential. Changes in science and policy, however, are discernable which make possible a new form of scientific activity (domain-based research). This, in conjunction with collaborative forms of social and technical innovation, capable of social amplification, could offset the regressive tendency.

A value confusion over the relations of welfare and development persists from the culture of industrialism in which they have been in conflict. This is impeding the transition process in which they are congruent. The 'systems' and 'ideo-existential' aspects of this reversal are treated in chapter 10, the historical and contemporary in chapter 11. The changes taking place are of an order as great as those which in Neolithic times first produced large-scale societies.

The post-industrial society is structurally present but culturally absent (chapter 12). Emerging values which may be adaptive under Type 4 conditions are identified (chapter 13). These recall pre-industrial and re-define industrial values in a new gestalt.

Such values are unlikely to establish themselves unless a new social context emerges through the spread of trans-bureaucratic organizations and the creation of a common 'ground' through the influence of the media of the information technology (chapter 14).

Chapter 15 focusses on the practical difficulties faced by the individual and his family during the transition to post-industrialism. Emerging life styles are creating changes in the 'risk-security balance' which increase 'the burden of choice'. Large numbers of people who will pass in the next two or three decades into relative affluence are unprepared for this challenge. A new art of living will have to be learned. The chances of this being accomplished depend on the appropriateness of the interdependent systems of personal values, organizational forms and modes of political regulation which emerge. We have some though not complete power to influence the outcome.

A new culture of politics is required which, assisted by 'adaptive planning', is able to regulate complex, rapidly but unevenly changing societies–based on the acceptance of pluralism and the surrender of power (chapter 16).

The project assignments on which Part Two is based have been carried out in a North American context, in the United States and Canada, while those which gave rise to Part One took place in a European context, in Britain, Ireland and Norway. While this has given us a trans-Atlantic perspective, opportunities to undertake work in the developing countries would have widened our outlook in a way most desirable for social scientists concerned with environmental processes. Short visits and participation in international seminars are no substitute for the experience of undertaking projects in other parts of the world.

As regards our general orientation we would describe this as contextualist in the sense intended by C. S. Pepper (1950). As social psychologists we have been profoundly influenced by the field-theoretical approach of Kurt Lewin, from which it was but a short step to develop a 'systems' orientation. Because our natural tendency is to think ecologically we have tended to work with members of other disciplines and to be committed to the inter-disciplinary development of social science.

Though one of us is now in the United States and the other in Australia this does not prevent us from continuing to work jointly. The final revisions for this book were made at the Management and Behavioral Science Center of the Wharton School of the University of Pennsylvania, where Dr. Russell Ackoff has become a 'co-producer' of much that we have been involved in during the last few years, whether in Britain or the United States.

Two books have recently appeared whose approach is closely related to our line of thought but which were not available when this was being developed. They are *Future Shock* by Irvin Toffler, Random House, N.Y. (1970) and *Beyond the Stable State* by Don Schon, Shaw & Temple, London (1971)–based on his Reith Lectures.

Also *Towards a New Management Philosophy* by Paul Hill, Gower, London (1971) is being published this autumn. It is about the Shell Philosophy project in which we took part and which is mentioned several times in the text.

We are grateful to a fellow social ecologist, Sir Geoffrey Vickers, for writing the foreword. His influence will be apparent from our sub-title.

January 1972 *F.E.E. & E.L.T.*

Part one

The next thirty years: concepts, methods and anticipations

F E Emery

Planning and a Conceptual Basis for Predicting the Future

Social Science Capabilities
The Relevant Time Horizon for Social Science Planning
The Active Role
Directive Correlation
Planning as the Extension of Choice
Summary
The Problem of 'The Present'
Temporal Gestalten
Overlapping Temporal Gestalten
Four Factors Influencing the Predictability of Temporal Gestalten

We are here concerned with identifying the needs which the social sciences should be prepared to meet in the next thirty years.

There is a common feeling that men's needs for understanding and controlling themselves and their societies may, in the next thirty years, be different from their current needs. It is difficult to deny the validity of these feelings. Practically all of our social institutions, the regulative ones of family, church and state, as well as the productive ones, have been evolving in this century at a rate which promises further substantial, qualitative changes in the next thirty years. Certainly significant changes are to be expected in the ways men can relate themselves to others, and in the way they are jointly confronted by their environment.

Social Science Capabilities

This is a challenge to the social sciences. Their capabilities are in understandings, scientists, methods and, not least, institutionalized arrangements for teaching, research and for relating the social sciences to the society. None of these capabilities can be quickly grown, run down, redirected or coalesced. Together with the intense competition with other sciences, professions, etc., for rare resources, the social sciences have their own theoretical blinkers, vested professional interests and institutional rigidities.

Apart from the fads and fashions with which we are still afflicted, our recent history suggests that we cannot expect an important new insight seriously to affect the growth or direction of social sciences in under five years. Major projects like those of *The Authoritarian Personality,* or Bruner's *Studies in Cognitive Growth,* take at least five years from inception, through research and publication, to widespread impact on the research teaching and applications of others. Institutional growth and professional training almost certainly require us to think in terms of more than five years to get from inception to self-sustaining growth. However, this scale of from five to ten years is not enough to guide effectively current decisions on investment.

The Relevant Time Horizon for Social Science Planning

Within the time scale of from five to ten years, one could hope to plan for the development of important concrete capabilities, but the existence of capabilities (adequate theories, methods, personnel and organization) exerts a significant effect on what is expected of social science. The planners must consider therefore not just such questions as 'Will these resources create this capability?' but also 'How will the emergence of these capabilities transform the environment for which they are planned?'

This second question is neither fanciful nor trivial. We have ample historical evidence to show how theoretical and institutional advances in the social sciences have, willynilly, attenuated or amplified the demand for other contemporary resources. It is hard to avoid the conclusion that each wave of planning must seek to create the conditions required for successfully planning the next wave, i.e. for wise investment in any period one needs to have some image of the character of the next period and sufficient notions about the third period to sense what might be the purposes of the second period.

In the social sciences this would seem to involve a foresight of twenty to thirty years, but not a detailed forecast of this whole period. Decisions must be made with regard to current resources but there is no suggestion in this model of preempting later decisions—rather the opposite, that later decision-makers should be at least as well placed, as far as one can foresee, to make the choices they will wish to make.

The Active Role

Given planning, the social sciences can play an active role in the next decades, not simply a passive one—they can seek to modify their social environment so that men can better pursue the ends they desire and not be left to adapt blindly to whatever emerges. If the social sciences were concerned simply to adapt to the next thirty years, then planning for the future would be based on extrapolations of the sort that 'by the 1990s x proportion of the population of size X will be in schools; given the past rate of increase in educational psychologists per ten thousand students, we must plan for a supply of such-and-such numbers of these specialists'. This approach would leave unconsidered whether it might not, for instance, be better to develop a theory of pedagogy or a re-organization of industrial culture that would radically change the multiplier effects of the educational psychologist or the pre-eminence of schools as places of learning.

Paradoxically, the problems of making predictions would be easier if the social sciences stuck to a passive role. By actively seeking to augment man's ability to control himself and his institutions, the social sciences are more likely to enhance genuine unpredictable novelty. Men would have greater control, but the manner in which they would exercise it would be less obvious than if they continued as at present.

We have suggested that the approach to the next thirty years is very much influenced by whether one assumes for the social sciences an active role or a passive one. The concept of planning for a real world entails an active role; it is not reducible to predictions or forecasting (Drucker, 1957). But Jerome Bruner in his presidential address to the Society for the Psychological Study of Social Issues (1964) made the essential point that the active role is not that of dictating:

> . . . however able we are as psychologists, it is not our function to decide upon educational goals . . . . The psychologist is the scouting party of the political process where education is concerned. He can and must provide the full range of alternatives to challenge the society to choice (pp. 23-23).

Given the stress being laid on the distinction between active and passive roles and the possibilities there are for misinterpretation, it is probably desirable to spell out the conceptual distinction.

Directive Correlation

The distinction we have been trying to make has been sought after by Sommerhoff (1950, 1969) in terms of 'adaptation' and 'directive correlation'. Adaptation refers to the responses available for dealing with emergent environmental circumstances. The concept of directive correlation encompasses adaptation in that it allows for that system of causal relations in which the environment is actively influenced to determine the kinds of responses that will subsequently be adaptive.

The relation between these two concepts of adaptation and directive correlation can be stated precisely in diagrammatic form:

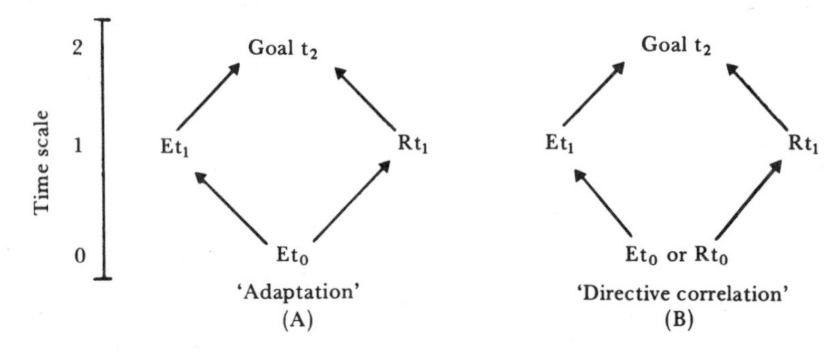

FIGURE 1

Both of these diagrams depict causal processes linking environmental conditions and (system) responses occurring together at time t_1 with their state at an earlier time t_0 and with their resultant end-state or goal condition at a later time t_2.

Both of these diagrams allow for variation in:

(a) the range of initial conditions of both the system and its environment. It is understood that these conditions may vary independently of each other and do not constitute a closed system within the initial state of one would determine the initial state of the other;
(b) the range of environmental conditions at t_1 for which is a corresponding range of responses;
(c) the degree of matching of these, as manifested in the probability or precision of achieving the goal; and lastly
(d) the time scale represented by the back-reference period, $t_1 - t_0$ and the attainment period, $t_2 - t_1$.

The causal models depicted in the diagrams differ in one critical respect. Diagram A defining adaptation is restricted to the effects of variation in initial conditions of an environmental nature, that is, it represents a stimulus-response relation. This, we hasten to add, is not a simple cause-effect relation. As Angyal phrases it:

> The external influence in such a case is not the mechanical cause of the reaction of the organism, but is the *stimulus* which *prompts* the response. The response is mainly determined by the intrinsic tendencies of the organism . . . . The external condition which prompts the response acts not by its properties as such but by its properties relative to the organism. The stimulus quality of an object is a relational property, that is, one which the object derives from its relation to the organism (Angyal, 1941, p. 36).

The stimulus is, with respect to the organism, embedded in and predictive of heteronomous environmental processes, ones that take place independently of anything the organism does. An object or event in the environment has stimulus qualities only insofar as it couples together such separate environmental and system processes.

'Adaptation', however, represents only the passive form of directive correlation. The other is the active form of coupling that occurs, for instance, when a man lights a fire. In this case, his wit and action sets off an environmental process that enables him by appropriate responses to pursue goals of warmth, cooking, of visual contact, of security, of distillation etc. Making fires is not only an adaptive response to the sun going down but can be a starting condition for *purposive* activity. (Since first writing this (1967) it has become clear that the distinction being made between the passive and active forms of directive correlation can be more rigorously defined as the distinction between goal-directed systems and purposeful systems, Ackoff and Emery (1972). Goal-directed systems can achieve the same end-state by a different course of action in each of two or more environments: purposeful systems can also achieve the same end-state by *choice* between two or more courses of action in the same environment.)

To be applied to the next thirty years of the social sciences, this simple model of directive correlation would have to be elaborated because:

(a) the key environmental processes are people who are capable of directively correlating their activities with the social sciences;
(b) in any real situation the social sciences will be involved in more than one other process; and
(c) the time scale involves a hierarchy of directive correlations within which the goals of the earlier ones are the starting conditions of the following.

The second and third elaboration do not affect the basic properties of the simple model, namely, that where a system can perceive, learn and choose, it is able to determine its future to a degree that is not possible for a system which relies on passively adapting.

Planning as the Extension of Choice

The first elaboration clarifies Bruner's assertion (and our belief) that the active role of the social sciences in the coming decades is not reconcilable with the social sciences seeking to determine the future of man. Unlike the other sciences, the social sciences cannot be indifferent to their subject matter. They cannot, in fact, expect to survive, let alone grow, unless they pursue ends that are shared by their chosen objects of study. No matter how cunning or devious the social scientist becomes, it is almost certain that his subject matter would eventually outmanoeuvre him, as no physical particle could. This is not a new observation: 'Suppose the phsiognomist ever did have men in his grasp, it would merely require a courageous resolution on man's part to make himself again incomprehensible for centuries.' (Lichtenberg, 1788, quoted by Hegel, 1949, p. 345.)

The survival and growth of social science presupposes a role in which it enhances the range and degree of directive correlations that men can form between themselves and their environment. Specifically, this might mean increasing the range of relevant conditions that men can take into account, increasing the range and efficiency of the responses they are able to make or extending men's awareness of the goals they might successfully pursue. In each of these ways the social sciences can contribute to men's ability to choose and to make the next thirty years.

This contribution is meaningful if, in fact, men have some ability and some desire to shape the future. We assume this to be the case, allowing only that:

(a) men can proceed from the objective conditions of the present;
(b) they tend to pursue only those goals that seem to be achievable (and hence may often be blind to possibilities that have newly emerged); and
(c) the means they choose may frequently have unanticipated consequences for other goals.

Our search for a conceptual basis for predicting the future led us to the concept of directive correlation in relation to taking the active role. Before we take up the nature of the present and of temporal gestalten we will summarize our arguments concerning the relations of planning and choice.

(a) there is a need for developments in the social sciences that go beyond their present concerns;
(b) this development needs planning;
(c) the planning needs to be in a context of expected social developments for several decades ahead;
(d) the planning should be more than projection or forecasting; it should be premissed on those properties of men and their institutions that enable them to actively adapt to their environments to make choices;

(e) planning should actively seek to extend the choices men can make, not to dictate them.

The Problem of 'The Present'

Even if we agree about what ought to be done by way of planning, we are no further advanced with respect to:

(a) knowing how to detect social developments several decades ahead or
(b) knowing what developments we should actually plan for.

In the remainder of this chapter we will examine some of the concepts and methods that might help us to determine the shape of the future. After this we can tackle the main question of what future.

A prediction of the future can always be challenged by pointing out that we can only know what we have experienced or are experiencing—that the future does not yet exist and hence cannot be experienced, cannot be known. This scepticism reduces itself to the position that we can know only what is presently experienced because the past is also non-existent and we have no way of experiencing and hence knowing whether what we *think* was experienced was actually experienced. These objections cannot be allowed to rest there. To be consistent one has to define what is the 'present', and if one insists that past and future do not exist and hence cannot be known then the present becomes the split second of immediate experience and knowledge; knowers and knowables disappear.

This attitude to prediction is no more useful to understanding what we actually do than is the other Laplacean extreme which suggests that the past and future are completely given in the present array of matter and energy. Our own experience of successful and unsuccessful prediction is a far better guide to what we might be able to achieve in trying to assess the future requirements for the social sciences. Granting the compelling point that we cannot experience that which does not exist we are still prepared to agree that we know something scientifically if we know we could, given present conditions, create the relevant experiences (by experiment, test or observation). This copes not only with why we believe that we know something of the past, but also with why we believe we know something about the future, e.g. we can experimentally demonstrate that exposure to present conditions will lead to a particular set of events at some point in the future. At a trivial level we can say that, given the numbers taking up sun-bathing today, there will probably be many more with sunburn tomorrow.

These latter considerations give us good reason for rejecting a sceptical viewpoint about prediction and accepting the question more usually asked by Everyman 'How do you know that?'–allowing that only under some special circumstances will he ask 'How *can* you know that?' However, we have in our reposte implicitly redefined the notion of present. The present within which we can potentially carry out a confirmatory experiment or collect the ingredients of sunburn is not the immediate conscious present of the sceptic. Is this simply a sleight of hand or are there other grounds for redefining the notion of 'present', apart from the fact that 'the immediate present' is an impossible useless concept?

This problem was brought to a head in psychology with Lewin's concept of contemporaneous causation as applied to the life space of an individual. Lewin, and subsequently Chein, suggested that, just as much of the present is organized into spatial gestalten, so the present is embedded in 'overlapping temporal gestalten'.

Temporal Gestalten

The experience of a melody presupposes experience of a temporal gestalten. A sneeze can be part of the present but so is middle-age part of the present of a middle-aged person and the nineteen-seventies part of the present of a railway organization. Any person or group is at any instant in many 'presents', each corresponding to what is a phase of the temporal gestalten in which he or it is embedded. In dealing with living systems, whether species, population groups or individuals, we have been led to the viewpoint that there are laws corresponding to the whole course of a living process. This is because we have identified in these processes parts which coexist throughout the duration of the process, and in their mutual interaction and interdependence generate the causal relations characteristic of that process.

Certain (not all) of the characteristics of events arising within a process, or the emergence of phases of a process will be determined and hence can be predicted by the laws governing that process. However, by the same reasoning, the phases will possess certain characteristics of their own arising from the mutual determination of their sub-parts; and hence laws of their own. These characteristics will not be determined by the characteristics of the preceding phases unless these arise from laws of the total process and except in so far as the preceding phases determine the starting point of the phase in question.

Sommerhoff has stated these propositions in a more rigorous and exact way in his concepts of long-, medium- and short-term directive correlations (corresponding to phylogenetic, adaptive learning and behavioural responses) and of the hierarchies that can arise between them. For our purposes, it is enough to note that it is consistent with the principle of contemporaneous causation to regard certain types of past and present events as causally related to and predictive of events that have yet to occur or to be experienced. These are the events that arise in the course of the process and are mutually determined by the laws that govern that process. In psychology, for example, the facts of maturation and learning are of this type. The pre-requisite for prediction is a knowledge of the developmental laws. In the absence of this knowledge even the meaning of the immediately present facts cannot be understood; we can even regard it as being theoretically impossible to gain this understanding by knowing all about every immediately present fact. (This is the problem of Laplace's super-mathematician and the illusion of some super computer schemes for integrated data systems.) In addition to a knowledge of the laws governing different classes of living processes, we need a knowledge of earlier facts if we are to know how those laws are operating in a specific individual process, and hence to know the effects they will have on later phases.

Overlapping Temporal Gestalten

So far we have considered only the case of a single process ('temporal gestalt', system or 'directive correlation') and its parts, and have implied that the whole burden of causation is within a process. This is, of course, a travesty of reality. Many of the phenomena we observe arise from the interaction of processes that we are unable to treat as if they were parts of a more inclusive process. *When such independent processes overlap, a new process emerges and a class of events is generated which has no history prior to the beginning of the interaction.* There are clearly degrees of independence. The interpersonal life that will emerge in the marriage of a man and a woman from the same culture is probably more predictable than that which would emerge if they came from different cultures. In any case, these hybrid processes seem to entail a special degree of unpredictability. The sufficient conditions for these newly emerged classes of events cannot be found in the prior history of the individual processes.

Our main suggestions about the theoretical possibilities and limits for prediction can be spelt out more clearly with reference to simple diagrams. Throughout, we will be concerned with predicting the future of concrete individual processes (e.g. of the U.K. or of John Smith). We will not be considering how one builds up predictive knowledge for a class of repeated or repeatable processes, nor will we consider forecasting techniques for processes that display only quantitative change.

Let us assume that figures A, B and C in Figure 2 represent the scope and temporal extension of three living processes (which could, for instance, be ecological, social or psychological). Let t_0 represent the present and t––, t–, t+ and t++ represent past and future points in time.

Four Factors Influencing the Predictability of Temporal Gestalten

(a) **Familiarity.** In the situation represented in Figure 2, we would expect to be able to predict the state of A at t+ better than we could B at t+ (provided, of course, that A and B are the same kinds of system). *The general principle is simply that for any system there is a minimum number of its component positions that have to be filled by parts before the system is recognizable.* In practice, we do find that the more of its course a system has run, the easier it is to understand. On the same grounds we would regard C as unpredictable at t_0.

(b) **Phase Distance.** Figure 3 represents a situation where a and b are phases of A. While some prediction about the future part of a is theoretically possible, there is no basis for predicting the specific characteristics of phase b. Beyond phase a one could only make predictions of the kind discussed with reference to Figure 2, i.e. predictions about the more general features of system A.

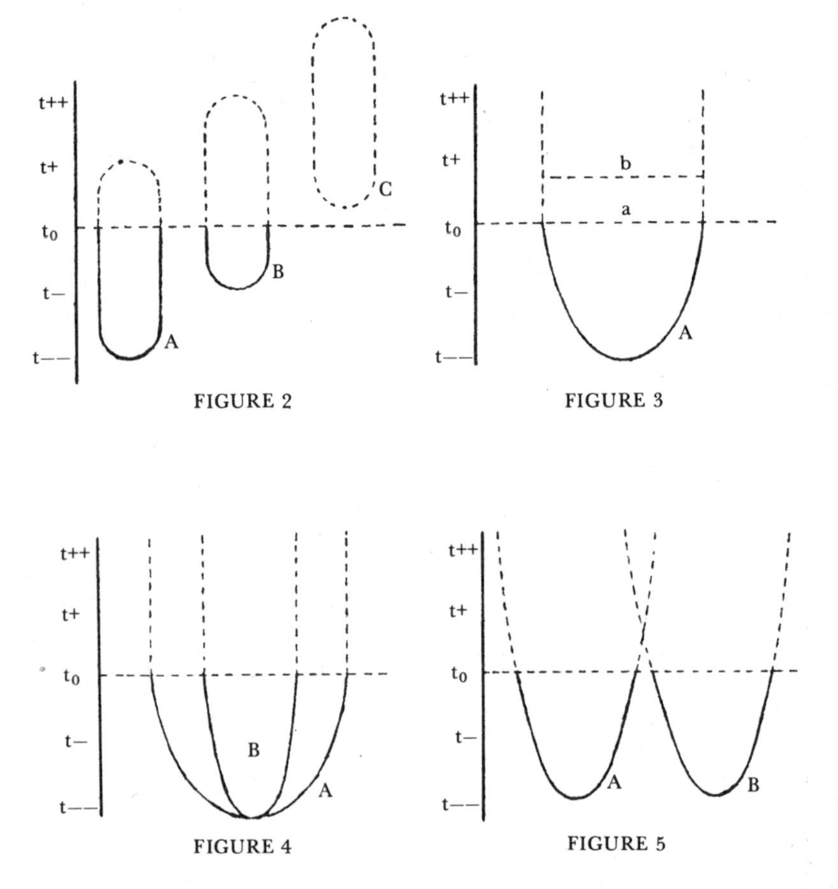

FIGURE 2

FIGURE 3

FIGURE 4

FIGURE 5

(c) **Inclusiveness.** Figure 4 represents a situation where A and B are coextensive in time but B is a part process of A. One would expect that predictions about A would be theoretically easier than predictions about B taken alone. The basis for this expectation is the general property of part-whole relations. A sets some of the parameters of B and hence, whatever one knows of the values likely to be taken by B, one knows more if one knows how these parameters might change. The future of B is dependent upon the future of A in a way that A is not dependent upon B. At the same time, predictions about A will be less specific than could be predictions about B.

(d) **Emergent Overlap.** In Figure 5 we have two processes which are presumed to interact after some point t+ in the future. If A and B survive the interaction, some of their system properties may predictably survive. What seems unpredictable are the processes set up by the interaction and the changes occurring in A and B if they become directively correlated to form larger inclusive system.

It would be too much to expect that the above mentioned situations constitute a complete set or that our interpretations are all equally defensible. It will suffice if we have made the point that there are genuine theoretical questions involved in predicting the future (as distinct from methodological ones) and if we have explicated our own assumptions. These assumptions will guide our search for appropriate methodologies and our strategy for identifying future changes.

Chapter 2

Complexity Reduction

Identifying the System and its Generating Function
Two Difficulties in Complex Reduction
Values as Indicators of System Tendencies
Analysis Based of the Leading Part

Identifying the System and its Generating Function

Given the conceptual model of overlapping temporal gestalten ('processes', 'systems' or 'directive correlations'), the general methodological problem is clear—to identify 'the constructive principle' (ends or 'focal condition') that characterizes the system or sub-system whose development we are trying to predict. A good methodology will be one that enables us to predict earlier.

There are two aspects to this methodological problem.

(a) *To identify the system* in terms of its *components* and the *dimensions* in which they are arranged. This is not simply a matter of counting off those things that display a sufficient degree of interdependence to warrant being treated as a system. *Most systems, particularly in their early stages, are incomplete ('open gestalten')* and hence system identification can only be considered adequate if one can enumerate not only the present members and their relations but, from these, also the unfilled positions in the system and the strains they create. The notion of incompleteness is implied in statements of the sort that political system X or person Y is immature.

Under this aspect we can classify seven of the twelve modes of prediction that Bell identified as 'structural constraints' and 'requisites', and 'operational systems' and 'codes'. These identify major system characteristics and lead to predictions of persistence or decay. Also here is the mode of predicting from the 'overriding problem' (i.e. the goal of the system), 'the prime mover' (the basic starting conditions) and 'phase theories' (identifying a temporal hierarchy of goals and starting conditions).

(b) *To identify the characteristic generating functions* (c.g.f.) *of the system.* The underlying notion here is that insofar as a system generates its successive phases, it will exhibit some temporal series of behaviour which, if quantified, could be represented by a mathematical series. Such a mathematical series has the property that its characteristic generating function can be identified from a finite part of the series (even if the series is infinite), and given the *c.g.f.*, one can predict from any starting point the subsequent members of the series.

These two aspects of the problem are not always explicitly dealt with in published models of prediction. In the mode that Bell calls 'social physics', we have had many attempts to postulate the characteristic generating functions of identifiable systems: Marx's concept of the relations to the means of production provides a classical instance. However, in the mode of 'trends and forecasts', we typically find that the models deal with *aspects* of a system without explicitly relating these aspects to the behaviour of the total system. By taking into account more aspects, as in the current U.S. government move from national income accounting to 'social accounting', these models may move closer to predicting total system behaviour. This is particularly likely if, as in the quoted case, the selection of new aspects to measure is guided by explicit analysis of the system.

The remaining two of Bell's modes of prediction concern neither identification of system components and dimensions nor their characteristic generating functions. They are techniques that could be used to enrichen or test any of the other ten modes. Thus, construction of 'scenarios of alternative futures' is explicitly concerned with identifying the range of end-states that might occur and hence might be chosen as desirable. It is not concerned with identifying what current conditions might render some futures more probable than others. 'Decision theory' has some potentiality for determining the relative 'rationality' for a system of the different courses of action that might be predicated by any of the ten modes of prediction classified above.

Two Difficulties in Complexity Reduction

Bell's concern was with modes of predicting future states of *large complex social systems*. This is also our concern, so it is relevant to discuss several special difficulties that arise with studying these systems:

(a) their complexity is greater than that which we have so far learnt to cope with in our separate social sciences;
(b) the sharing of parts between different sub-systems is so great that their subordination to newly emerging processes can be difficult to detect—the parts appear to be still functioning as parts of the established familiar systems although perhaps a little more erratically.

The rest of this chapter will be focussed on possible solutions to the first difficulty.

It seems to us that the social sciences cannot hope to cope with the complexities of large complex social systems unless they take as their *unit of analysis* something larger than and inclusive of these systems. Specifically, the unit must include the system and its immediately relevant environment.

Lewin (1936b) and Ashby (1956) both offered general solutions to the problem of representing high orders of system complexity. Both took the system-environment as their unit of study and both recognized that reduction should be to those features that were genotypical, that is, the characteristic generating functions. These, hopefully, would emerge from use of topology or application of 'stability' theory to differential equations. In developing their general solutions both found it difficult to avoid reduction of their unit of study to just another large complex system. Lewin's psychological system tended to become encapsulated within the 'psychological life space' and Ashby's system tended to become just a part of a more comprehensive system.

That their efforts proved to be premature should not blind us to the daring nature of their proposed solution and its essential correctness. To reduce the complexity of our systems models they proposed to add in the even greater complexity of the environment, with the further complication that the two sets of processes tend to demand incommensurate models drawn from many different disciplines!

What makes sense of this apparently foolhardy approach is the fact that if reduction is not to yield a misleading caricature of a system's future it must be a reduction to genotypical characteristics. In living systems the most fundamental genotypical characteristics are the system-environment relations that determine survival (i.e. continued living and reproduction). In populations of living systems capable of active adaptation each system is part of the environment of the others and they constitute together *a social field.* As one actively adapting system A becomes sensitive to the potentially critical part that another can play in a given setting, so it becomes sensitive to how B might have acted in earlier settings to create the conditions with which it is now faced; how B might set in later settings to undo or build on what A now does; how some other system, C, might act if A's relation with B takes a certain course. The multiple short- and medium-term directive correlations that thus emerge constitute an extended field of directive correlations–a social field. Such social fields have properties that persist in the absence of any one of their constituent systems and at the same time determine critical conditions for adaptation and survival of these systems. The social field includes less than the total environment of a system but as a first approximation it offers more hope of advancing the program of Lewin and Ashby. Subsequent approximations would probably seek to relate the social field to the biological and physical environment, e.g. in seeking to make predictions with respect to population growth, resource utilization, pollution. However, even for these successive approximations the same methodological point is relevant. Prediction of the future states of human populations requires that the unit of analysis includes the social field and that reduction should still retain at least the core dimensions of the relation between the two.

If we take this social field as a special kind of superordinate system then there are two questions that should guide the search for complexity reduction in a complex social system:

(a) What are the system-environment relations that typically determine survival in this social field?

(b) What are the system tendencies toward generating such relations?

These questions need to be jointly answered. Out attempt to provide some kind of answer is reported in chapters 4, 5 and 6. For the moment we wish to outline three less general approaches that have largely been concerned with methods for answering question (b):

i. values as indicators of system tendencies;
ii. starting conditions as indicators;
iii. analysis based on the leading part.

Values as Indicators of System Tendencies

The demands for survival in a particular environment should place value on certain kinds of preparatory behaviour at the expense of others; changes in the conditions of survival should induce changes in these values or goals. The direct study of what is valued should therefore enrich the predictions that could be made from study of survival conditions alone. Several specific methodologies have been suggested for studying values.

Churchman and Ackoff (1949) have argued that where we have a reason to believe that something, X, is a value for a social system, we can test this belief by seeing whether there has been, over an appropriately long period of time:

(a) a tendency to increase the efficiency of the means for pursuing X;
(b) a tendency to greater use of the more efficient means;
(c) an increased conscious desire to achieve X.

(a) and (b) could be otherwise formulated as an increase in the range and degree of the directive correlation having X as a goal. (c) is a necessary condition because both (a) and (b) could be manifested by a process that arises from the accidental overlap of two temporal gestalten (as in Figure 5 above). In the case of warfare we can certainly see an increase in the efficiency of weapons and a marked tendency for their usage to spread, even to warring Congo tribesmen. The absence of condition (c) gives some grounds for doubting whether the wholesale murder of others is a basic social value. If a, b and c are all present as in pursuit of health in modern human society we have very good reason for believing that it is a value for that society.

As a methodology, the Churchman-Ackoff proposal seems a particularly promising start. The most desirable elaboration of this method is probably that which will help order the relation between values, e.g. for what values are people prepared to sacrifice most. One can readily envisage how this method might help us predict the longer term shifts in value that plague the 'trend and forecast' people.

A more popular methodological approach to the same problem is provided by the combination of sample surveys and value tests (e.g. Cantril, 1965). This is essentially limited to part (c) of the Churchman-Ackoff model and hence, for use as a basis for prediction presupposes some evidence about (a) and (b). Without the latter, one cannot be sure whether the support for a value is, over the long run, declining, stable or growing. A growing conscious desire for religion might for instance reflect concern over the decline in religious institutions.

The second methodological approach to complexity at this level of generality is that of *identifying the starting conditions* (coenetic variables) that have arisen from the past adaptive responses and act as a constraining and guiding influence on subsequent preparative behaviour. This has appeared to be the really scientific approach during those past generations when the value oriented actions of men so frequently produced unanticipated and undesirable consequences (e.g. World War I and economic depressions). One marked attraction has been the appearance of a social system in which the vast complexity of past individual contributions has been congealed and crystallized into a much fewer number of formal organizations. This is particularly clear in the economic field. The state of these organizations at any one time seems to be a firm basis for what will subsequently emerge. Combined with a variant of the first method–analysis of the values pursued by organizations–it seems particularly attractive. However, the individual orientations left out of this organizational approach may well nullify this attempt to reduce complexity. Developments like those of Nazi Germany suggest that these 'residual' non-organized behaviours are an important condition for what will emerge in a society (Cantril, 1941; Fromm, 1950).

Both the study of values and of starting conditions appear in practice to achieve less than the necessary reduction in complexity. Almond's 1960 model, for instance, would require repeated sampling of several hundred sub-samples of organizations. Similarly, the range of values that can conceivably be supported in a human population is excessively large.

Analysis Based on the Leading Part

This brings us to suggest that there is a methodology intermediate in generality between the Lewin-Ashby model and the two just discussed. The intermediate one concerns the notion of *'the leading part'*. In this case, the reduction is not, as it were, a reduction across the board to pick out a key element present in all of the parts. Selecting the leading part seeks to reduce the total complexity by ignoring a great deal of the specific characteristics of all but one part. At its extremes we have the reduction to a *figure-ground* relation in which the leading part is considered in relation to all the other parts taken together as its ground (the 'internal environment' of the total system). Throughout this range of possibilities *the method is basically that of establishing which part it is whose goals tend to be subserved by the goals of the other parts or whose goal achievements at t_0 tend to determine the goal achievement of all the parts at $t+$.*

Practical use of the methods of value study or structural analysis usually involves an implicit assumption about what is the leading part, e.g. McClelland's 1961 study of achievement values as a driving force in modern history and the Marxist mode of production theory. The values of the elite or the character of a central organization (or set of like organizations) can readily form the basis for predictions about the future. There would be a better basis for prediction if the intermediate step of selecting the leading part has an explicit methodological basis. One expected windfall from asking 'What part acts as the leading part?' is that major phase changes might be identified. Most studies of developmental phases in individuals or societies seem to identify a change in phase with a change in the leading part.

These suggested methodologies do not add up to an established discipline for study of the large complex so-called 'socio-economic-technological systems'.[1] They do indicate that this order of complexity is not an insurmountable barrier and that some progress has already been made.

[1] The phrase 'socio-economic-technological systems' has been used in M.I.T. contributions to the Year 2000 Commission. It appears to derive from the Tavistock Institute of Human Relations' use of 'socio-technical systems' (Trist, 1951). For our part, we think it misleading. A society is composed of *substantive* socio-psychological organizations and socio-technical organizations, and at the same time is a population or aggregate of concrete individuals. There are economic, political and affective *aspects* which can be *abstractly* identified in all human organizations. Thus we will discuss social and technical systems, we will not discuss economic systems. This does not preclude discussion of economic aspects.

Chapter 3

The Early Detection of Emergent Processes

Concealment and Parasitism
Shared Parts
Phases in the State of Competing Systems
The Sigmoid Character of the Growth Process
Changes in the State of Symbolic Systems
Symbol Analysis
Value Analysis and Linguistic Usage

Concealment and Parasitism

The second major difficulty in predicting the future states of large complex social systems, that of early identification of emergent processes, poses far more perplexing methodological problems. However, if social life is properly characterized in terms of overlapping temporal gestalten, then many of those processes that will be critical in the future are already in existence in the present. If this were not the case, it would be difficult to see *how such processes could quickly enough muster the potency to be critical* in the next thirty years. Thus, for instance, the conditions for World War I were laid before the end of the 19th Century, and correctly perceived around 1900 by such oddly gifted men as Frederick Engels and the Polish banker, Bloch (Liddell Hart, 1944, p. 26).

An obvious question must be asked at this point: 'Is this not the same class of evidence that is the basis for extrapolative prediction?' Such evidence does include some evidence of this class, but its most important additional inclusion is of *processes that are not recognized for what they are.* The early stages of a sycamore or a cancer are not obviously very different from a host of other things whose potential spatio-temporal span is very much less; likewise with many processes in social life.

One suspects that the important social processes typically emerge like this. They start small, they grow and only then do people realize that their world has changed and that this process exists with characteristics of its own. Granted that there are genuine emergent processes (otherwise why worry about the next thirty years), then we must accept real limitations upon what we can predict and also accept that *we have to live for some time with the future before we recognize it as such.*

Yet it is not simply foolhardy to think that we may enable ourselves more readily to recognize the future in its embryonic form. There are almost certainly some regularities about these emergent phases. Social processes which, in their maturity, are going to consume significant portions of men's energies are almost bound to have a lusty growth. They do not, by definition, command human resources at this stage, and hence their energy requirements must be met *parasitically,* i.e. *they must in this phase appear to be something else.* This is the major reason why the key emergents are typically unrecognized for what they are while other less demanding novel processes are quickly seen. A social process which passes for what it is not should theoretically be distinguishable both in its energy and informational aspects. Because it is a growing process, its energy requirements will be substantially greater (relative to what it appears to do) than the energy requirements of the maturer process which it apes. Because it is not what it appears to be, *the process will stretch or distort the meanings and usage of the vocabulary* which it has appropriated. The energy requirements may be difficult to detect not only because we lack scales for many of the forms of psychic and social energy, but also because a new process may in fact be able to do as much as it claims (e.g. TV to amuse), but do it so much more easily as to be able also to meet its own special growth requirements. The aberrations of linguistic usage are, on the other hand, there to see.

In trying to go further along these lines, we will first try to explain why there are probably significant although undetectable processes operating in the present. The explanation we will give itself suggests some methodologies that might aid early detection. For reasons of continuity we discuss these before tackling the logically prior question of whether there is any particular reason for trying to achieve early detection.

Shared Parts

Complex social systems like the human body rely a great deal on the *sharing of parts.* Just as the mouth is shared by the sub-systems for breathing, eating, speaking, etc., so individuals and organizations act as parts for a multiplicity of social systems. Just as there are physiological switching mechanisms to prevent us choking too often over our food, so there are social mechanisms to prevent us having too many Charlie Chaplins dashing out of factories to tighten up buttons on women's dresses (in *Modern Times*). I think that *it is this sharing of parts that enables social processes to grow for quite long periods without detection.* If they could grow only by subordinating parts entirely to themselves then they would be readily detectable. If, however, *their parts continue to play traditional roles* in the existing familiar systems, then detection becomes difficult indeed. The examples that most readily come to mind are the pathological ones of cancer and incipient psychoses. Perhaps this is because we strive so hard to detect them. In any case, healthy changes in physical maturation, personality growth or social growth typically follows the same course. Once we are confronted with a new fully-fledged system, we find that we can usually trace its roots well back into a past where it was unrecognized for what it was.

Phases in the State of Competing Systems

If this is, in fact, the reason for most or even some important social processes being undetected, then it suggests methodological approaches. Despite the redundancy of functions that the parts tend to have with respect to the role they play in any one sub-system, one must expect some interference in the existing systems as a new one grows. Angyal, from his analysis of competing psychological systems within an individual, has suggested a general classification that could serve as a basis for analysing social systems (Angyal, 1966). This is as follows:

1. When the emerging system is relatively very weak, it will tend to manifest itself only in the parasitical effects it has on the energies of the host system—in *symptoms of debility.* The host system will find it increasingly difficult to mobilize energy (people) for their functions and there will be a slowing

down of their responsiveness to new demands. The balance of forces may oscillate so that these symptoms occur in waves and make the functioning of the existing social systems less predictable.

At any time a social system experiences a fair amount of uncontrolled variance (error) in its operations. The reasons for an increase in this variance, of the kind we are discussing now, will typically be sought for inside the system itself, and measures may be taken to tighten up its integration. The unpredictable oscillatory effects are likely to encourage a wave of experimentation with new modes of system functioning. All these symptoms have behavioural manifestations and are hence open to study. The methodological strategy of operational research is that of proceeding via analysis of the variance of systems and this would seem particularly appropriate here.

2. When the emerging system is stronger but still not strong enough to displace the existing system, we can expect to see *symptoms of intrusion.* What breaks through are social phenomena, like the swarming of adolescents at the English seaside resort of Margate several years ago, which are clearly not just errors in the functioning of the existing systems. At the same time, because of the relative weakness of the emerging social systems, they will usually only break through because they have short-circuited or distorted the functioning of the existing systems. Their appearance will not obviously reveal the shape of the emerging system. However, if we are aware of the possibility that these phenomena can arise from emerging systems, it should not be beyond our ingenuity to develop appropriate analytical methods (as has been done in psychology for detecting from slips of the tongue the existence of competing psychological systems).

3. When the emerging system has grown to be roughly in balance with the existing systems there may be *mutual invasion.* At this stage it should be obvious that there is a newly emerging system but mutual retardation and the general ambivalence and lack of decisiveness may still lead the new system to be seen simply as a negation of the existing system. The methodological task is to identify, in the chaotic intermingling of the systems, characteristics of the new system which are not simply an opposition to the old. Once again we find that this is not an entirely new methodological problem for the social scientist.

The Lewinians gave considerable attention to this in their studies of *'overlapping situations'*, for instance adolescence, when new and old psychological situations are frequently invading each other. Barker, Wright and Gonick (1946) specified five dimensions that they found helpful to sort out what was being done to what, by what. These dimensions are consonance, potency, valence, barriers and extent of sharing of parts.

(a) **Relative consonance.** Two or more overlapping situations requiring behaviour from the system that is more or less congruent. The degree of consonance ranges from *identity,* where the same behaviour meets both situations, through *consonant* where different behaviours are required but they are non-interfering, *interfering,* to *antagonistic.*

(b) **Potency.** The influence of one situation relative to all simultaneously acting situations.

(c) **Valence.** The relative desirability or undesirability of the situations.

(d) **Barriers.** The relative difficulties confronting the system as it tries to make progress in the different situations.

(e) **Extent of common parts.**

With the aid of these dimensions, they were able to spell out many of the behavioural properties of invading systems. These conceptual dimensions have been sufficiently well defined to permit ready translation into other theoretical schemes.

The fact that early detection may be possible does not in itself make it worthwhile pursuing. The fact that early detection increases the range of responses and hence the degree of control a system has over its development does make us interested. There are facts about the growth of social change that suggest that each unit step in the lowering of the detection level will yield a disproportionately greater increase in the time available for response. Put another way, it would yield a disproportionately richer projection of the future from any given time.

The Sigmoid Character of the Growth Process

The next points I wish to make by referring to Figure 6. Let lines A and B represent two courses of growth over time. If social processes typically grew in the way represented by curve A, then we might well feel that early detection was not a pressing problem. At this steady rate of growth, we might expect that when the scale got to the level of ready detection (D on the vertical axis) we would still have the time C – A (horizontal axis) in which to aid, prepare for or prevent the new system getting to critical size (c on vertical axis). All of this is simple enough and the assumptions do not seem unreasonable because so many of the changes in the physical world and in our physical resources do grow in something like this manner.

In fact, a great many of the growth processes in social systems appear to be more like that represented by curve B. These growth curves are common enough in all living populations (and some physical systems) where each part has

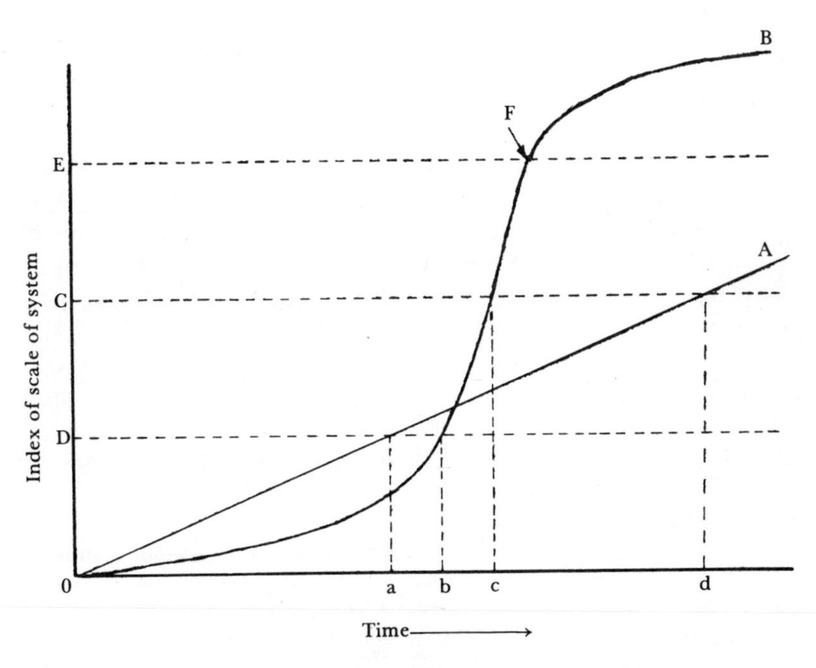

FIGURE 6

powers of multiple *replication,* but in this case we are primarily concerned with *recruitment* of existing parts to a new social system. What appears to contribute most to the prevalence of type B growth curves in social systems is the fact that these possess *the property of highly developed symbolical communication.* What is absent (because it is past, distant, or as yet only anticipated) can be represented by one part to the other parts. Their mutual co-ordination and regulation is vastly extended, and so is, as a result, the contagion of changes. One important implication of this is that a new system may, after a long period of slow and undetectable growth in the interstices of the society, suddenly burgeon forth at a rate which produces a numbing effect on the society, or at least drastically reduces the range of responses to it. The general notion may be explicated by again referring to Figure 6. If the point of critical size is somwhere near where I have marked in C, then it is in the nature of the type B curve that there will be less time between detection and critical size than would occur with a type A growth curve: i.e. $T(c - b) < T(c - a)$.

Although, in this section, I have concentrated on the early detection of emerging systems, the present line of argument has implications for the fate of rapidly growing systems. The sort of growth that occurs between detection at point D and point E on Figure 6 can only too easily be seen as a type A growth. Even if the growth up to point D is reconstructed, the curve O to E may be seen as a pure exponential growth curve which will continue on at an increasing rate of growth towards point F. This has been well illustrated by Price (1961). Bringing together statistical evidence on the growth of science he shows that it has the characteristic of the curve O to E. This characteristic seems to underly the recent scientific ethos that the sky is the limit for scientific growth. However, he argues that the next stage of growth will be like the curve F to B, not a continuation of the curve from the intersect with D to F.

It is indeed apparent that the process to which we have become accustomed during the past few centuries is not a permanent feature of our world . . . . The normal expansion of science that we have grown up with is such that it demands each year a large place in our lives and a larger share of our resources. Eventually that demand must reach a state where it cannot be satisfied, a state where the civilization is saturated with science (p. 113).

For science in the United States, the accurate growth figures show that only about *thirty years* must elapse between the period when some few per cent of difficulty is felt and the time when the trouble has become so acute that it cannot possibly be satisfied . . . we are currently in a period in which the onset of a manpower shortage is beginning to be felt . . . (115-6).

To this I can only add the obvious point that the method of study proposed by Price should include our preceding proposals. The decline in growth rate may occur not only because there is a limited supply of recruitable parts, but also because new systems are competing for existing parts.

Once again we find that elucidating the general nature of social changes is a fruitful way of identifying methodologies for furthering our ability to predict change in individual social systems or processes. The sigmoid type of growth curve (i.e. our B curve) has been a potent tool in the study of all types of living systems.

Changes in the State of Symbolic Systems

There remains a further general class of methodologies for early detection. These derive in the first place from the fact that man is not just a symbol-user in the way we have been discussing. His fundamental relation to his environment is a symbolical relation.

. . . the function of the so-called mental processes is essentially a semantic one. By this we mean that 'psychological contents' function as symbols and the psychological processes are operations with these symbols (p. 56). In the psychological realm life takes place, not through the interaction of the concrete individual with a concrete environment—which is only tangential—but by the interaction of symbols representing the individual and the environment (Angyal, 1941, p. 77).

As Tomkins has argued, our present knowledge of man suggests that if our perception mirrors nature, our consciousness is a mirroring of this mirror by the conceptual ordering of our memories.

. . . afferent sensory information is not directly transformed into a conscious report. What is consciously perceived is *imagery* created by the organism itself. The world we perceive is a dream we learn to have from a script we have not written . . . . Instead of putting the mirror to nature we are . . . putting the mirror to the mirror (Tomkins, 1962, Vol. 1, p. 13).

The essential adaptive advantage is that the error inherent in this process makes learning possible. For our purposes the relevance is that man's responses are to the world as he symbolizes it and not directly to the world as it presents itself to his eyes, ears, etc.

In the second place, while this mechanism of consciousness (awareness of awareness) is a condition for learning, the learning itself is not conscious (certainly not necessarily conscious). Thus man's symbolical representation of the world may change to represent changes in that world without his being conscious of the change. In so far as he is unaware of these changes they may remain unrecognized, or, if manifested in his behaviour, be puzzling, trivialized, or segregated parts of his projected world picture.

I have dwelt on these properties of the individual human being because they are basic to any joint human activity whatever the scale or complexity. On available evidence, it would seem that men live and have always lived in a cultural world which is created and maintained by the symbolic transformation of the actual world and the imputation or projection thereon of the meaning and values by and for which we live. My second point about individuals seems also to hold for social systems, namely that the social symbols, the myths, beliefs, values, language, fads and fashions change without any necessary awareness of what the change means or to what they correspond. More concisely, there can be awareness of world changes without awareness of that awareness; and this awareness can be manifested in man's communicative behaviour as well as in his other behaviours. When these manifestations are recorded in oral traditions, in art forms or writing, it is theoretically possible that analysis of the records will reveal the existence of social processes which existed at the time, were sensed and lived with but not consciously grasped. At least three methodologies of different levels of generality have begun to emerge here. For convenience we label them as follows: (a) symbol analysis; (b) value analysis; (c) analysis of linguistic usage.

Symbol Analysis

We use the term symbol analysis to refer primarily to the methods of Jung and his followers. On the same assumption that basic changes in the life conditions of large groups may be detected in symbolic changes, Bion has speculated that we might be able to develop a method of inferring such basic changes from statistical fluctuations in psychosomatic symptoms (as unconscious individual symbolizations) and in the value of money (as in part reflecting aggregate psychological valuation) (Bion, 1961, pp. 105-113). This approach cannot be ruled out. The ethnologists and ecologists have together shown the nearly ubiquitous nature of symbols in living populations and their contribution to the natural selection of populations. Since this, it has been difficult to write off the possibility that the human species might have evolved through the use and selection of some similar innate cognitive programmes involving 'perceptual concepts' (Arnheim, 1954).

Less tentatively, we can accept the possibility that cracks and repairs in man's umbrella of symbols might well presage the obvious emergence of major social processes by a long period of time. Neumann (1966), Marcuse (1956) and McLuhan (1964) have made much of the notion that signs of our present condition were present in the painters, poets and writers of fifty years ago. As might be expected, McLuhan is particularly outspoken on this. He quotes Wyndham Lewis as writing: 'The artist is always engaged in writing a detailed history of the future because he is the only person aware of the nature of the present.' To this he adds his own judgement, that 'the artist is the man in any field, scientific or humanistic, who grasps the implications of his actions and of new knowledge in his own time' (McLuhan, 1964, p. 65). For these reasons, McLuhan sees his own method of detecting the future in the present as an application of the analytical techniques of modern art criticism. Just because these methods are esoteric, we cannot afford to ignore them.

Value Analysis and Linguistic Usage

The analysis of values has already been touched upon because this, like the analysis of symbols and linguistic usage, offers a radical reduction in the complexities with which we would have to deal. In each of these we would be using men themselves as a filter of what is important.

The analysis of linguistic usage is at one level a commonsense way of sensing the way a person is developing or the way a people are tending to go. The very way in which people are speaking about things is often a valid indication of changes in the way they are looking at the world, even though they insist in all honesty that they have in no way changed their views. This method is a basic ingredient of psychiatric practice. At the social level, it has been applied to the content analyses of films, women's magazines, etc., and, more intuitively, to tracing out the subtle shifts in the meanings of key concepts like 'work', 'leisure' and 'justice' (Arendt, 1958). Marcuse has given us a profound analysis of the relation between experience and linguistic usage. He sets the methodological goal of linguistic analysis as that of 'analysing ordinary language in really controversial areas, recognizing muddled thinking where it *seems* to be least muddled, uncovering the falsehood in so much normal and clear usage. Then linguistic analysis would attain the level on which the specific societal processes which shape and limit the universe of discourse become visible and understandable.' (Marcuse, 1964, p. 195). Drawing upon the empirical study of Karl Kraus, he specifies some of the features of the method:

For such an analysis, the meaning of a term or form demands its development in a multi-dimensional universe, where any expressed meaning partakes of several interrelated, overlapping and antagonistic 'systems'. For example, it belongs:

(a) to an individual project, i.e. the specific communication (a newspaper article, a speech) made at a specific occasion for a specific purpose;
(b) to an established supra-individual system of ideas, values and objectives of which the individual project partakes;
(c) to a particular society which itself integrates different and even conflicting individual and supra-individual projects' (Marcuse, 1964, pp. 196-7).

It will be noted that these are methods of gathering information about the different levels of system competition which we presented as the general model for early detection.

I mentioned earlier that these methods offered a reduction in the complexity which had to be coped with, because men will, if acting unwittingly, tend to symbolize the relevant changes and filter out for themselves the relevant changes. If acting consciously, they will typically see things through the ideologies of their times. This is, however, only a relative reduction. A profound reduction may occur with a Blake or Joyce. However, this may be of little use. How do we recognize a Blake or Joyce in our midst or understand what they are saying when they probably don't understand themselves? If these methods of analysis are to be effective, we shall still have to deal with samples of data that are very complex relative to our current analytical tools. It has been recognized that modern computers may bring us within reach of the point where the predictions of such highly perceptive individuals as McLuhan, Marcuse and Neumann can be converted to testable hypotheses. Stone's (1966) General Inquirer programme is a step in this direction, but it would still be necessary to identify the kind of system which one suspects is emerging. In other words, these methods complement the perceptive intuitive minds.

An example may illustrate and draw together some of the methods I have discussed under the heading of 'early detection'. It is desirable, of course, that we concentrate upon the general principles, not the concrete features of the example. Assume, for instance, that a resurgence of Nazism is thought to be likely in a given country. Early detection is desired in order to allow counteraction and yet it is expected that any such embryonic movement would actively seek to avoid detection until it had recruited enough strength to challenge existing social systems and overcome the conceivable counteraction. The recruitment of any particular individual can be hidden because recruitment does not entail total subordination to the party. The recruit can still continue to function as civil servant, waiter, husband, etc., although there may well be some falling off in the enthusiasm with which he now carries out his duties or even some change in the way he conducts them. However, even if each recruit in turn recruits several others each year, the growth rate, while sigmoidal, would put off the achievement of critical mass in a large nation for a long time (and of course increase the probability of detection). Therefore, in a large nation, a resurgent Nazi party would need to use the mass media. (Clandestine leaflets, papers and wall slogans would intensify efforts at detection.) They would have to penetrate and use the media in a covert way in order to avoid detection. However, to use it all they would have to shape the media content and style so that it propounded their *Weltanschauung.* It is not impossible to do this and at the same time avoid detection and counteraction. The aim would be to reach and to nurture the thinking of like-minded persons and these will tend to be more sensitive to low intensity messages than all but perhaps the obsessed anti-Nazis. Secondly, people can learn from a large number of trivial cues without being aware of just what led to the learning. This latter counts heavily against the obsessed anti-Nazi. He may well come to the firm conclusion that a particular medium has Nazi flavour and yet be unable to put his finger on anything that constitutes evidence for demanding counteraction.

In this case, how would the methods of symbolical analysis help to test hypotheses about the emergence of such a concealed symptom? Briefly, they would involve some sampling of media content because of the sheer mass of material going through them. The sample, if it were to be at all sensitive, would have to be handled by computers. The computer programme would need to be so designed that it could detect metaphors of the sort that Jung thought central to Nazi thinking, values of the sort that McGranahan (1946) found to distinguish the Nazi Youth from the U.S. Boy Scouts, and the more complex problems of syntax, grammar, vocabulary and even typography which Kraus found so revealing. For practical purposes the last would have to be restricted to the controversial political universe of discourse where in any case the effects are more significant. By repeating the study over time it should, theoretically, be possible to determine whether there is an embryonic growing process, more than one centre of growth or simply unrepentant, unburied remnants. It should not be impossible to go beyond mere detection to inferring structural properties and system orientations that differ from assumptions based on past experience.

As an example this is not entirely satisfactory. The hypothetical social process is conscious of its ends, consciously striving to use the symbolic processes of the society and consciously seeking to avoid detection. The latter does not simply cancel out the first two features to make it equivalent to a non-self-conscious social growth. Hence, although in this case symbolical analysis can only be usefully employed when the weaker system is strong enough to start intruding, it does not argue against symbolical analysis at the earlier stage when all that is present are symptoms of pressure.

Summary of Methodologies Discussed in Chapters 2 and 3

In this section I have outlined the following:

1. Two aspects of the general methodological problem:
   (a) to identify the system in terms of its members and the dimensions in which they are arranged;
   (b) to identify the characteristic generating function of the system.
2. Special methodological difficulties that arise with predicting the future of large complex social systems:
   (a) complexity;
   (b) early detection.
3. Methods that have been developed or proposed for overcoming these difficulties:

(a) *Complexity*

1. Ashby's model for studying conditions for survival;
2. Models for studying subordinate goals (values), e.g. Ackoff-Churchman, Cantril;
3. Models for studying the starting conditions for change;
4. Method of identifying 'the leading part'.

(b) *Early detection*

1. Model derived from the properties of weakly competing systems;
2. Sigmoid growth model;
3. Models based on analysis of symbols, values and linguistic usage.

Chapter 4

The General Characteristics of Social Fields—Environmental Levels

Western Societies as the leading Part
The Strategy of Overall Characterization
Placid, Clustered Environment
Disturbed Reactive Environment
The Turbulent Environment

Western Societies as the Leading Part

It was argued in the first section of this paper that the future will be largely shaped by the choices men make, or fail to make, and it will not be moulded simply by technical forces; it was argued that processes existing in the present can reveal some of the basic choices that will confront men over the next thirty years; and, finally, it was argued that social science should consider not only the provision of tools (trained personnel, institutions, theories and methods) but also the more active role of helping men to extend their visions.

On this basis I shall seek to identify current developments which are changing the conditions within which men can make their future, and shall look at these both in terms of the challenges they pose and the opportunities they create for further human development. This should reveal the areas within which growth in social scientific knowledge and capabilities can most help men to help themselves.

Following the conceptual scheme outlined in chapter 2, I shall move from consideration of the broader social systems to narrower ones. Following my own judgement, I shall start from consideration of the total social field of entities such as the U.K. and U.S.A. i.e., modern Western nations. I am assuming that within the inclusive social field constituted by the world population these currently constitute the leading part. The method of approach will be basically that of trying to answer the two basic questions of system-field relation posed in chapter 2.

Next, I shall assume that the current leading part in such systems is the productive system—the complex of interrelated socio-technical organizations concerned with the social (not household) production of material goods and services. For reasons given in chapter 3, I think that this method of proceeding is preferable to abstracting common phenotypical characteristic aspects such as the political beliefs or values. The next step follows the same procedure of identifying the information and communication industries as the leading part of the productive system. Because this last step puts us at two removed from the social field of modern Western nations I will then go back to see what effect this elaboration of the production system has on the total system.

Lastly, I shall touch upon the major boundary conditions of our primary unit. These appear to be (a) the relation of the modern Western nations to the more inclusive international field, (b) the biological inputs of these social fields, and (c) the natural resources upon which they rely.

Throughout, the concern will be with matters on which the development of the social sciences might have a bearing.

The Strategy of Overall Characterization

As pointed out in chapter 2, if there are predictions to be made about complex systems they are most likely to be valid if they are derived from analysis of the genotypical characteristics of broader social fields. This is, of course, only a theoretical point: we might have little or no information on which to assess the larger systems. This is, in fact, the reason for starting with the more limited strategy of choosing the Western nations as the leading part, although it is evident that they are part of a larger social field. Nevertheless, I do not wish to be like the drunk in the story who knew he had lost his watch up the dark alley but searched under the street lamp because there he had lots of light. There is a body of evidence accumulating about the growth characteristics of the Western type of society. This evidence is not of the sort that readily permits of graphical or mathematical extrapolation but it seems to permit of genotypical analysis. I will devote most attention to this analysis because it provides the framework within which more detailed predictions of part processes can be made. A simplified version of this analysis has been published (Emery and Trist, 1965), but so much weight is being placed upon drawing inferences that the argument should be spelt out more fully.

In trying to characterize large complex social systems, I am reminded that some behaviours of both organisms and organizations are a function of gross overall characteristics of their environment (Chein, 1954). We can advance our knowledge of these behaviours if we can identify the properties that best characterize the overall environment and the system behaviours necessary for adapting to them.

This is not a new strategy for the social sciences (see chapter 2). Thus, in psychology the Lewinians were able to demonstrate the lawful behaviour of 'human-beings-in-cognitively-unstructured-situations' and of 'human-beings-in-overlapping-situations' (Barker, 1946). A great deal of so-called learning theory is of the same kind, that is, the study of behaviour in overly simplified 'conditioning' situations, structured 'meaningful' situations, complexly structured or 'problem' situations, overly complex or 'puzzle' situations. Similarly, Chein (1954) has pointed to the gain that may be had for psychology from the study of environments that in overall terms are relatively stimulating or stimulus lacking, relatively rich or poor in goals or noxiants, cues or goal paths, easy to move in or sticky, etc.

In the field of economic organization, a similar scientific strategy has yielded the characterization of markets as classical competitive, imperfectly competitive, oligopolic, monopolistic. These again are attempts to define ideal types of overall environments and again have been relatively successful in showing the lawfulness of some of the behaviour of economic enterprises.

In the field of military organization, the great post-war disputes over optimum size of operating units, optimum weapon capabilities for size of unit, optimum organization of support facilities and command structures have all centred on the problem of the changes in the more general characteristics of the battlefield environment following the advent of tactical nuclear weapons.

The solution we sought appeared to be along these lines. Therefore we concentrated on those dimensions of the environment which constitute its *causal texture* (Emery and Trist, 1965). By causal texture we meant, following Pepper, Tolman and Brunswick, the extent and manner in which the variables relevant to the constituent systems and their inter-relations are, independently of any particular system, causally related or interwoven with each other.

For simplicity of exposition we considered the relevant variables only as goal objects or noxiants for the constituent systems (i.e. having different relative values for the systems with the values ranging from positive to negative) and assumed that there is some sense in which these can be spoken of as more or less distant from or available to the organization and hence requiring more or less organizational effort to attain or avoid. Already, it will be noted, something has to be known about the organization in order to delimit the environment in this way. We have to know what is of potential value to it and what are possible courses of action.

For our purposes, we found it necessary to distinguish only four levels of organization of environments.[1]

Placid Random Environment

The simplest form of environment is that in which goals and noxiants are distributed randomly and independently through the environment. That is, a placid, random environment, placid because of heteronomous processes in the environment of the system.[2] This ideal corresponds to Simon's 'surface over which it (an organism) can locomote. Most of the surface is perfectly

[1] Any attempt to conceptualize a higher order of environmental complexity would probably involve us in notions similar to vortical processes. We have not pursued this because we cannot conceive of adaptation occurring in such fields. Edgar Allen Poe did go into this problem in his short story *'Into the Vortex'*. He intuited that there was a survival tactic if drawn into a whirlpool—namely to emulate an inanimate object. To strive in one's own way was to perish. Folk lore and natural history are full of similar lessons about 'playing possum', 'playing dead'. For our purposes we are inclined to regard these as survival tactics rather than adaptive *behaviour.* In case there may be something to the hunch that a type V environment has the dynamics of a vortex it is worthwhile noting that vortices develop at system boundaries when one system is moving or evolving very fast relative to the other—like a Watt County, L.A., and between the developed and underdeveloped countries.

[2] We certainly do not wish to convey the meanings associated, for example, with 'placid tranquillity'. An old-fashioned mad-house or a Nazi concentration camp could constitute the kind of environment we are defining. Our choice of the term was largely dictated by our need for a contrast with the environmental disturbances that characterize some of the more textured environments.

bare, but at isolated, widely scattered points, there are little heaps of food'. ... '... the food heaps are distributed randomly' (Simon, 1956, pp. 129-38). It also corresponds to Ashby's limiting case of 'no connection between the environmental parts' (1960, S,15/4); to Toda's 'Taros Crater' (1962, p. 169); and Schutzenberger's stochastic environment (1954, pp. 97-102). The economists' classical market comes close to this ideal environment. Thus, although this represents an extreme type of environment, there has been wide recognition of the need to postulate it as a theoretical limit. The relevance goes deeper than simply providing a theoretical bench-mark. This low level of organization may frequently occur as the relevant environment for some secondary aspect of an organization and is also quite likely to occur in humanly designed environments for the reason that such simplified environments offer maximum probability of predicting and controlling human behaviour, e.g. Adler's 'Sociology of the Concentration Camps' and the experimental environments of conditioning theory.

The survival of an organization in a placid random environment is a fairly simple function of the availability of these environmental relevancies and the approach-avoidance tactics available to the system (i.e. its response capabilities). So long as the environment retains this random character, it does not make much difference if there is more than one need and it is not necessary to postulate any complex organizational capacity for identifying marginal utilities or substitution criteria. 'We can go further, and assert that a primitive choice mechanism is adequate to take advantage of important economies, if they exist, which are derivable from the interdependence of the activities involved in satisfying the different needs' (Simon, 1956, p. 134). We can go even further and assert that in the absence of differences in relative value—the nearest goal object being the best—system behaviour in those environments does not involve choice.

Given that some environments with which we are concerned may be judged to be 'random, placid' in their causal texturing we may still be concerned with the effects of different degrees of randomness on survival.

An indication of the importance of these differences is given by Simon's consideration of the effects of 'range of vison'.

Increased range of vison in a random environment is equivalent to increasing the area of non-random environment immediately surrounding the system. From the computations he makes for his very clean model of the random environment '... we see that the organism's modest capacity to perform purposive acts *(sic)* over a short planning horizon permits it to survive easily in an environment where random behaviour would lead to rapid extinction. A simple computation shows that its perceptual powers multiply by a factor of 880 the average speed with which it discovers food' (Simon, 1956).

It is not enough just to characterize an environment and postulate minimum survival characteristics for systems in those environments. Environment and system do not just co-exist side by side. They interact to the point of mutual inter-penetration. Some aspects of the environment become 'internalized' by the system and some aspects of the system become externalized to become features of the environment. There are three modes of inter-relation that we will consider for each level of environment:

(1) instrumentality;
(2) planning;
(3) learning.

The first two are forms of interpenetration of the system into its environment. What Trist and I labelled as L_{12} relations; where L_{11} relations represent potentially lawful processes within a system, L_{22} lawful processes within the environment and L_{21}, L_{12} the influences from environment to system and system to environment, respectively (Emery and Trist, 1965). The third mode, learning, may and, hopefully, usually does manifest itself in the use of instruments and planning. However, I think we come closer to the core of meaning if we look at learning as an interpenetration of the environment into the system (i.e., as an L_{21} relation). Viewed in this way the main concerns of learning theory should be with (a) the informational structure of different environments, (b) the kinds of behaviour in these environments that justify the title of 'learning behaviour'.

Again we turn to Simon, this time for a good consideration of the relevance of *instrumentality* in placid, random environments. This appears in his treatment of 'storage capacity'. Storage capacity is not a response capability but it may be instrumental in extending or restricting response capabilities. In an example dealing with organisms it is natural to think of storage capacity as intrinsic, like a stomach or fatty tissue, but in thinking about systems this is an irrelevant assumption—parts of the environment may just as readily be used by the system for storage purposes. Simon found that storage capacity was a highly relevant parameter of survival although not as much so as reduction of randomness. One would expect therefore to find that systems exposed to these environments would tend to hoard, and to find that this was adaptive.

The appropriate *planning* mode in this environment has been stated very precisely by Schutzenberger, namely that under this condition of random distribution there is no distinction between tactics and strategy, and 'we find that *the optimal strategy is just the simple tactic of attempting to do one's best on a purely local basis*' (Schutzenberger, 1954). This aptly describes the marketing approach of successful traders in Petticoat Lane and other such flea-markets. They do not know where their next mark (customer) is coming from and when he shows up their only concern is with closing the sale to their maximum advantage, without thought to subsequent consumer satisfaction.

Ashby has suggested that the best tactic in the circumstances can be learnt only on a trial and error basis and only for a particular class of local environmental variances (Ashby, 1960, p. 197). I agree that the circumstances place peculiar restrictions on learning but not with the suggestion that the appropriate learning behaviour is trial-and-error. The experimental environments in which trial-and-error has been observed to be adaptive have been complexly joined environments—puzzle situations for the organisms concerned (Thorndike, 1911, Hamilton, 1967). The classical learning situation most closely corresponding to the placid, random environment is that devised by Pavlov and his followers. In this situation sound-proofing, restrictive harness for the animal etc. are used to create a blank unvarying environment and the animal is exposed to random encounters (that is, random for the animal) with some goal objects and some other specific stimuli which are unrelated to the goal objects, except for co-occurrence in space and time (Pavlov, 1928). It would be difficult to devise a better reproduction of a random, placid environment. *The learning 'behaviour' observed is conditioning* not trial-and-error. Strictly speaking there is no behaviour involved as there is no element of goal-seeking, the system is just conditioned. Theoretically the probability of survival should improve as the system is conditioned to take advantage of any departures from randomness in its environment.

A final point is that higher order systems must accept the degrading of their learning to simple conditioning, their strategies to simply following their noses etc., if they are to survive in these placid environments. However, as Zener found with his dogs (Zener, 1937) and Alder found in reviewing human behaviour in concentration camps, higher order systems will strive to utilize any elements of non-randomness to create more order and permit themselves to perform closer to their level.

Placid Clustered Environment

More textured, but still essentially a placid environment is that which can be characterized in terms of clustering of the goals and noxiants. The goals and noxiants occur together in space and time with varying probabilities that are potentially knowable for the system. This level of environmental texturing was introduced by our discussion of the significance for placid random environments of any reduction of randomness. It happens to be the kind of environment with which Tolman and Brunswick were concerned and corresponds to Ashby's serial environment and the non-stochastic environment within which Schutzenberger could identify the minimal characteristics of goal-directed behaviour.

The structuring that exists at this level of texturing enables some parts of it to act as signs (local representatives) of other parts or as means-objects (manipulanda, paths) with respect to approaching or avoiding. However, as Ashby has shown, survival is almost impossible if a system attempts to deal tactically with each environmental variance as it occurs or as it is signalled (signalling having the effect of multiplying greatly the density of confrontation) (1960, p. 199). Survival in environments of this kind requires a second-order of feedback involving some sort of threshold mechanism so that reaction is evoked less readily and only to the more general aspects of the environment—to the clustering which will reveal itself only through a manifold of particular occurrences.

This is the critical feature of adaptation to this kind of environment, namely that choice of strategies emerges as distinctively more adaptive than choice of tactics. It no longer follows that 'a bird in the hand is worth two in the bush'. To pursue the goal object that it can see, the goal object with which it is immediately confronted, may lead the system into parts of the field which are fraught with unforeseen difficulties. Similarly, avoiding a present difficulty may lead the system away from parts of the environment that are potentially rewarding. Adaptation of these environments therefore requires as a minimum that a system be goal-directed (Sommerhoff, 1969)—that for each of a number of different concrete situations it has a course of action that is determined more by the goal it pursues than by the immediate presenting of goals and noxiants.

In this sort of environment, it becomes possible to seek a best strategy where optimality is limited only by restrictions upon knowledge. Survival of a system becomes conditional upon its knowledge of its environment. In the extreme case, if enough is known of the structure of the environment so that 'the map's projection has been changed to that of the really optimal matrix, the distinction between strategy and tactic *(again)* disappears' (Schutzenberger, 1954). This differs from the randomized environment in that here strategy tends to absorb tactics. Given the omnipotence of a Laplace the tactics could be derivable from the strategy.

The objective of a system in this type of environment also has certain characteristics. In the placid, random it could have none, apart from tactical improvement and hoarding against a rainy day. In this second type the relevant objective is that of *'optimal location'*. Given that the environment is non-randomly arranged, some positions can be discerned as potentially richer than others, and the survival probability will be critically dependent upon getting to those positions. So much of management of organizations is concerned with planning, that it is worth considering some of the approximations that are appropriate in this type of environment:

(i) Domain selection. The recognition of clustering itself so that, at the level of strategic planning, one is concerned with relatively few clusters which can be approximately characterized as units instead of with a multitude of individual objects. This lowers the cost of information gathering and processing;

(ii) the development of a *hierarchy of strategies* as in the rules for trouble-shooting in complex equipment; in this way the off-putting effects of the unexpected occurring may be buffered by modifying lower order strategies but retaining intact higher order ones.

(iii) the *assignment of step functions* to the values of goals and noxiants instead of trying to act on a continuous range of values; the average human being, for instance, tends to break a continuum into five steps (Jordan, 1968, p. 137).

(iv) the backward determination of the strategic path. This is by far the least demanding procedure once the strategic objective is selected (Schutzenberger, 1954). This, however, does require subsequent adjustments of the strategic objective to fit the available paths.

'Planning by approximation' may represent only that lowest level of planning which Ackoff calls 'satisficing' (Ackoff, 1970). Nevertheless it represents what is possible and adaptive, given the information structure of a placid clustered environment. By definition an environment is only of this type when the system is limited to information about the relative probabilities of co-occurrence of goals and noxiants.

Naturally the planning behaviour reflects the *learning behaviour* that is possible. Systems are no longer limited to mere conditioning. This was clearly demonstrated by Zener (1937) when he duplicated the classical conditioning experiments with one modification: he freed the dogs of much of the restraining harness so that they could react to textural characteristics of the experimental setting, other than just the conditioned and unconditioned stimuli. The learning behaviour he observed, and filmed, was not conditioned behaviour but *goal directed meaningful behaviour.* The major characteristics of this type of learning have been studied and formulated by Tolman:

(1) The organism brings to a problematic situation various

systematic modes of attack, based largely on prior experiences.

(2) The cognitive field is provisionally organized according to the hypotheses of the learner, the hypotheses which survive being those which best correspond with reality, that is, with the causal texture of the environment. These hypotheses or expectancies are confirmed by successes in goal achievement.

(3) A clearly established cognitive structure is available for use under altered conditions, as when a frequently used path is blocked.

The third relation we have been considering, *instrumentality*, is also different in this environment. In the placid random environment the instrumental relation seems to be limited to variety-reducing forms such as hoarding and hiding (reducing the effects of environmental variation in goals and noxiants respectively). In the placid, clustered environment there is evidence that systems can use parts of the environment to increase the variety of courses of action open to them, i.e. to use them as tools. The classical experimental demonstration of this is Kohler's work with apes (Kohler, 1927). As originally reported these studies suggested that the experimental situation presented such a richly textured environment for the apes that their successful learning to use tools must manifest a higher order than just meaningful learning. It seemed that apes were capable of purposeful problem solving with the insights into causal structures that that implies. However, the knowledge that has now accumulated about the innate response capabilities of apes makes it fairly certain that their tool using was in response to an environment which was for them only placid and clustered (Chance, 1960).

Disturbed Reactive Environment

The next distinguishable level of causal texturing is one that we have called the *disturbed-reactive environment*. It approximates to Ashby's ultrastable system and the economists' oligopolic market. In this we simply postulate a placid, clustered environment in which there is more than one system of the same kind, and hence the environment that is relevant to the survival of one is relevant to the survival of the other. Formally, one could postulate a placid random environment with more than one system present, but I do not think that co-presence makes any difference to the concepts one needs to explain what differences in randomness would occur in that environment (which might be why the social sciences have such difficulties in linking up with so called reinforcement theories of learning). Co-presence makes a real difference in a placid, *clustered* environment because the survival of the individual systems requires some strategy as well as tactics. In this environment, each system does not simply have to take account of the other when they meet at random, but it has to consider that, what it knows about the environment can be known by another. That part of the environment to which it wishes to move is probably, for the same reason, the part to which the other wants to move. Knowing this, they will wish to improve their own chances by hindering the other, and they will know that the other will not only wish to do likewise, but will know that they know this. In a word, the presence of others will imbricate some of the causal strands in the environment. The causal texture of the environment will, through the reactions of others, be partly determined by the intentions of the acting organization. However, the environment at large still provides a relatively stable ground for the arenas of system conflict. Because of this, conflicting systems 'regarded as a unit, will form a whole which is ultrastable' (Ashby, 1960, p. 209).

How can competing systems constitute a stable unit in a disturbed, reactive environment? Given the relatively static nature of the environment within which the competition occurs, then it is possible (as it was for the individual organization in a placid clustered environment) for strategies to evolve that limit the disruptive effects of competitive strategies or competitive tactics. One would expect these strategies to be broader and take longer to emerge than those needed in a placid, clustered environment. They would not, however, differ in principle.

By starting from consideration of the causal texture of the environment and the way information flows from this, we avoid the dilemma of the economists' models of imperfect competition, duopoly etc. As Ferguson and Pfoutts point out (1962), those models yield predictions of inherent instability despite the observable fact that stability is commonly achieved. Ferguson and Pfoutts show that stability can be deduced, however, if information flow and learning are taken into account. By taking into account environmental properties, we find, as Simon found with the simplest environment, that we have less need to inject into our systems models (or models of man) a host of special *ad hoc* mechanisms, and we are less likely to come to false predictions.

One could maintain that this sort of disturbed reactive environment makes no difference to the distinction between strategy and tactics that we made for placid, clustered environments. I am inclined to think that it does. If strategy is selecting the '*strategic objective*'—where one wishes to be at a future time—and tactics is selecting an immediate action from one's available repertoire, then there appears in these environments to be an intermediate level. One has not simply to make sequential choices of actions (tactical decisions) such that each handles the immediate situation and yet they hang together by each bringing one closer to the strategic objective; instead one has to choose actions that will draw off the other organizations in order that one may proceed. The new element is that of choosing not only your own best tactic, but also of choosing which of someone else's tactics you wish to invoke. Movement towards a strategic objective in these environments seems therefore to necessitate choice at an intermediate level—choice of an operation[3] of campaign in which are involved a planned series of tactical initiatives, predicted reactions by others and counteraction. At this level the adoptive system is not just the one that can produce the right tactic for the right occasion (i.e. the goal directed system) but one that can *choose* the appropriate tactic.

If one tries to identify the level of system that is adaptive to disturbed reactive environments the critical criteria is that a system must be able, in at least some situations, to choose between two or more tactical moves either of which could further its ends, i.e. it must be a *purposeful* system.

There seems little doubt that even the formulation of strategic objectives is influenced by this kind of environment. It is much less appropriate to define the objective in terms of

[3] Cf. the use by German and Soviet military theorists of the three levels—tactics-operations-strategy.

location in some relatively static and persisting environment. It is much more necessary to define the objective in terms of developing the *capacity or power* needed to be able to move more or less at will in the face of competitive challenge. In business this would probably make it necessary to define objectives in terms of profitability, not profit. This formulation has an advantage in this kind of environment, in that there can be a day-to-day feedback of information relevant to this objective. In the former case, the day-to-day feedback about approach to a given location (e.g. percentage of market) may be extremely misleading. It may conceal the fact that the competitor has made the going easy by conserving his strength for a later stage (e.g. preparing to introduce an improved product).

The factors in this kind of environment that make it desirable to formulate strategic objectives in power terms also give particular relevance to *strategies of absorption and parasitism.* It is one thing in a placid random environment if other things can be characterized as goals or noxiants—they are either absorbed for the temporary sustenance they afford, or else avoided because noxious. It is another thing in a disturbed, reactive environment when the other, itself a system, has to be absorbed or be absorbed into because it is potentially noxious—because it is a source of important but unpredictable variance.

So far, with respect to this level of environment, we have discussed neither learning nor instrumentality. In our discussion of planning it seemed clear that the strategic objective of maximizing power dictates a corresponding mode of 'planning for the best solution' (what Ackoff calls *'optimizing'*—'to do as well as possible', Ackoff, 1970). We can now ask, what kind of learning behaviour in this environment enables a system to do this sort of planning?

The first part of the answer lies, as might be expected, in the changed informational structure of the environment. A placid clustered environment yields only information of concomitance (probable co-occurrence of goals and of noxiants). In a disturbed-reactive environment, with its independent causal agents, it becomes possible to distinguish between what is system action and environmental response and what is environmental pressure and system response. In other words, it becomes possible for a system to 'learn' the causal patterning of its environment. One further step in learning seems critical in this environment. Given that the other systems can also learn the underlying causal patterning, and direct their behaviour accordingly, it is necessary to learn the possible and probable recombinations of the causal pattern. This I suggest is the sort of learning that is involved in chess and other such genuine exercises in problem-solving (De Groot, 1965; Wertheimer, 1959).

As regards instrumentality, it is enough to note that the adaptive distinctions between strategy, operations and tactics enable a system to use parts of the environment to change other parts to the status of tools: in other words, to act as tool makers. There seems to be no inherent restriction in these environments to elaborating such tools to the point where they are fully adaptive to placid clustered environments.

The Turbulent Environment

The most complexly textured environments in which adaptive behaviour is possible, as distinct from sheer survival tactics, are *'turbulent fields'*. These are environments in which there are dynamic processes arising from the field itself which create significant variances for the component systems. Like the disturbed reactive and unlike the placid random and placid clustered, they are *dynamic* environments. Unlike the disturbed reactive, we are postulating dynamic properties that arise not simply from the interaction of the systems, but also from the field itself.

There are undoubtedly important instances in which these dynamic field properties arise quite independently of the systems in the social field (as with some of the earth and water movements in mining). However, in the conceptual series we are here elaborating, most significance attaches to the case where the dynamic field processes emerge as an unplanned consequence of the actions of the constituent systems; that is, those environments that represent a transformation of disturbed reactive environments. Fairly simple examples of this may be seen in fishing and lumbering where competitive strategies, based on an assumption that the environment is static, may, by over-fishing and over-cutting, set off disastrous dynamic

processes in the fish and plant populations with the consequent destruction of all the competing social systems. We have recently become more aware of these processes through the intervention of the ecologists in problems of environmental pollution. It is not difficult to see that even more complex dynamic processes are triggered off in human populations.

There are four trends that have particularly contributed to the emergence of these turbulent environments. Before stating these, however, let me briefly state that *these fields are so complex, so richly textured, that it is difficult to see how individual systems can, by their own efforts, successfully adapt to them.* Strategic planning and collusion can no more ensure stability in these turbulent fields than can tactics in the clustered and reactive environments. If there are solutions, they lie elsewhere.

The four trends that have together contributed most to the emergence of dynamic field forces are:

(i) The growth, to meet disturbed reactive conditions of organizations and linked sets of organizations that are so large that their actions are persistent enough and strong enough to induce autochthonous processes in the environment (I am here postulating an effect similar to that of a company of soldiers marching in step over a bridge or the pulsating budgetary requirements of the U.S., Soviet military establishments).

(ii) The deepening interdependence between the economic and the other facets of the society. The growing size and relative importance of the individual units not only creates interdependence within their economic environment; it also produces interdependence between what consumers want and what they think can be produced, between the citizen as consumer, as producer, as inhabitant, and as a social and political entity. This greater interdependence, when matched with the indepedent increase in the power of other citizen roles means that economic organizations are increasingly enmeshed in public reaction and in legislation and public regulation of what they do or might think of doing. The consequences that flow from the actions of organizations lead off in ways that are unpredictable. In particular the emergence of active field forces (forces other than those stemming from the individual organizations or the similar organizations competing with it) means that the effects will not tend to fall off 'with the square of the distance from the source' but may at any point be amplified or attenuated beyond all expectation. As a case in point, thalidomide as a simple cure for morning sickness created a major crisis for the international pharmaceutical industry and initiated a radical redefinition of responsibilities in one of the relations between science and the society. Similarly, lines of action that are strongly pursued may find themselves unexpectedly attenuated by emergent field forces, e.g. the U.S. 'War against Poverty'.

For organizations, these changes mean primarily a gross increase in their area of *relevant uncertainty.*

(iii) The increasing reliance upon scientific research and development to achieve the capacity to meet competitive challenge (which capacity, we suggested, tends to become the strategic objective in disturbed reactive environments). This has the effect not only of increasing the rate of change, but of deepening the interdependence between organizations and their environments. Choices that once appeared to arise from the market place are now seen as being taken by the organization on behalf of the customer–they are seen as manipulators of desire or, as with thalidomide, sorcerers' apprentices. It is not hard to imagine an organization finishing up in the dock of public opinion because it chose a line of technical development that appeared to suit its own needs but eventually left the economy in the lurch. The same trend appears in fields of public policy-making where competition over the allocation of resources is increasingly conducted by means of scientific research and analysis.

(iv) The radical increase in the speed, scope and capacity of intra-species communication. Telegraph, telephone, radio, radar, television, gramophone, typewriter, linotype, camera, duplicator, Xerox, calculator, Hollerith, computer: these names register a century of change that continues in an explosive fashion. Parallel with these has been a very great increase in speed and ease of travel, so that recorded communications flow in greater bulk at greater speed, and even the recording of communications becomes short circuited as it becomes easier for managers, scientists and politicians etc. to fly to each other than to correspond. We may recall that Trotter (1916) in searching for the conditions underlying social reactivity in living populations, postulated only two critical conditions: (a) some special sensitivity to their own kind; (b) some intra-species communication system. The change that has taken place in intra-species communication is a greater mutation than if man had grown a second head. The consequences are a great increase in the information burden and a radical reduction in response time in the system—a reduction which is unaffected by distance. Reaction takes place almost before action is formed. Even simple servo-systems with these properties readily get entangled in erratic 'hunting' behaviours. As the information burden approaches 'overload' it invites, in fact demands, radical counter-measures which tend to be maladaptive and increasingly unpredictable.

We will probably find that these trends are only part of the picture. However, they are in themselves real enough and may explain why we feel that consideration of the turbulent fields is a matter of central importance and not just a theoretical exercise.

What is less clear is how our society can adapt to these conditions. Ashby very wisely counsels that there may not be a solution to this problem:

> As the system is made larger (and is richly joined), so does the time of adaptation tend to increase beyond all bounds of what is practical; in other words, the ultrastable system probably fails. But this failure does not discredit the ultrastable system, as a model of the brain for such an environment is one that is also likely to defeat the living brain (1960, p. 207).

However, as a biologist, Ashby offers us the consolation that: 'Examples of environments that are both rich, large and richly connected are not common, for our terrestial environment is widely characterized by being highly subdivided' (1960, p. 205). It is my belief that this sort of environment is, in fact, characteristic of the human condition: that in some areas of his living man has always had to contend with turbulent environments. What is true is that just as the central matching process of consciousness has evolved to help protect the human organisms from information overload (Tomkins, Vol. 1, p. 14), so has man evolved his symbolic cultures to provide a man-made environment of tolerable complexity. What is significant of our present era is the emergence of a degree of social organizational complexity and a rate of coalescence of previously segregated populations that defy our current efforts at symbolic reductionism. Larger and larger parts of the lives of more and more people are being lived in conditions of environmental turbulence.

Adaptation to Turbulent Environments

Three Maladaptive Defences
Superficiality
Segmentation
Dissociation

If the preceding analysis is correct, then the next thirty years (at least) will evolve around men's attempts to create social forms and ways of life that are adaptive to turbulent environments or which down-grade them to the less complex types of environments. As argued earlier (chapter 2), survival questions are basic ones and, insofar as our societies take on turbulent properties, survival of our current institutional forms is challenged and men inevitably turn to these questions. We will try to spell out some of the ways in which survival can be sought. We cannot predict which paths men will actually take nor the actual means they will evolve in order to travel along the paths they choose. What we do know is that the social sciences could influence this process insofar as they give men greater insight into what they want and provide an extended range of means whereby they can pursue desired ends.

Three Maladaptive Defences

The proliferation of turbulent environments, and hence the range of social activities and the proportion of the population exposed to such environments, has probably been increasing rapidly since the turn of the century. Certainly, large numbers were exposed to the apparently heteronomous properties of the two world wars, the Great Depression and the rash of punitive autocratic regimes. Mass communications, particularly TV have made an even wider proportion of the population aware of the turbulent nature of this environment. The historians might determine the course of this process if we could provide adequate operational definitions of the distinctions we have proposed. For our purposes it will suffice if we can determine the kinds of responses men might be expected to make when confronted with environments that are predominantly turbulent. It seems only natural that they will seek ways of reducing the turbulence to the point where their learnt responses to disturbed-reactive environments are again adaptive. Any generally effective way of doing so implies *segregation (dis-integration)* of the social field so that men have to cope with only a part or an aspect of that field. All such responses are forms of passive adaptation. They are triggered off by the environment. They are also essentially defence mechanisms in that they seek to negate, downgrade, the environmental texturing with which they are confronted.

Following Angyal (1941) we can distinguish three dimensions of system integration and hence identify the three possibilities open to passive adaptation:

(1) the *depth dimension* ranging from the superficial manifestations of the system to its deeper underlying determinants. In a social field, reductionism on this dimension appears as increasing *superficiality*. This is achieved by denying the reality of the deeper roots of humanity that bind social fields together and on a personal level denying the reality of their own psyche.
(2) the *means-end dimension* of hierarchically ordered purposes, goals and sub-goals. Reductionism on this dimension appears as segmentation: sub-goals become goals in their own right and various goals are pursued independently of any over-riding purposes. To all intents and purposes the social field is transformed into a set of social fields each integrated in itself but poorly integrated with each other.
(3) the *transverse dimension* of co-ordination and regulation. Reductionism here appears as *dissociation:* the average constituent member of the social field reduces the degree of his association with others. In particular this is a reduction in willingness to co-ordinate one's behaviours with others or to allow one's actions to be regulated by the behaviour of others.

The distinctions being made here are of the greatest importance if social sciences are to come to grips with what is currently happening in the modern Western nations and if they are to assess the possibilities for active adaptation. However, whilst Angyal's distinctions are intuitively clear it cannot be proved in his theoretical framework that they constitute an exhaustive and mutually inclusive set of possibilities for maladaptive responses in a social field. Neither do his definitions constitute ideal operational definitions. Hence it is important to our purposes that distinctions based upon the rigorous formulation of purposeful systems (Ackoff and Emery, 1972) yield the same dimensions.

Thus, these passive adaptive behaviours are the behaviours of systems that are adaptive to disturbed reactive environments—purposeful systems. The purposeful choice of co-production with others will determine the quality of the social field. Where the choice is to reduce the complexity of the social field it must still manifest itself as a change on one of the dimensions of choice:

(a) probability of choice (familiarity; social or psychological distance);
(b) probable effectiveness (knowledge);
(c) relative value of intention leading to choice.

Reduction with respect to (c) yields *'superficiality'*—it does not matter much what the motives are for doing something (nor do motives provide an excuse for someone else being non-conformist). The characteristic attitude is intolerance, they wish that an increasing number of people just did not exist.

Reduction with respect to (a) yields *'segmentation'*, an enhancement of ingroup-outgroup feelings as people tend to simplify their choices by over-emphasizing the distances dividing people at all of the 'natural' dividing lines in the social field. The characteristic attitude is prejudice—'people ought to stay in their place'.

Simplification of choice on dimension (b), knowledge, is achieved by denying that what others do or could do as co-producers would enhance what one could do if guided by selfishness. Its effect is *dissociation* and anomie. The characteristic attitudes are those of indifference and cynicism.

Superficiality

This is achieved by denying the deeper roots of common humanity that bind social fields together and on a personal level by denying the deeper psychological roots of one's own behaviours. Traditionally this has been achieved by widespread use of the mechanisms of psychological repression and social suppression and oppression. The social manifestations of these modes were stereotypy of social behaviour, widespread symptoms of debility and unpredictable violent invasions of the suppressed (Cohn, 1957, and Pareto's residues, 1935).

This does not seem to be the current mode. Any and every possible source of human and social needs and behaviour is publicly explored. The dominant mode at present seems to involve some form of trivialization—if anything might lead to anything, then the motive for choice becomes pretty irrelevant and one chooses the familiar or the efficient. The dynamics have been clearly spelt out in Thorndike's puzzle box experiment. When a situation becomes too complex for organized meaningful learning, an organism regresses to *vicarious* trial and error behaviour—it responds first to this and then to that in a way which is unrelated to the structure of the environment but may be highly correlated with its own prejudices. Where this becomes a prevalent mode of adaptation, one may still get highly intelligent behaviour in the sense that an intelligence test measures the efficiency of responses to a strictly defined and limited set of starting conditions. Creativity will tend to be absent because creativity is essentially sensing or intuiting that a situation involves a different set of underlying determinants to those that are usually assumed. A most significant manifestation is outer directedness (Riesman, 1958). When responses are no longer critically and selectively related to hypotheses about their deeper psychological determinants, they no longer manifest such hypotheses and no longer challenge alternative hypotheses. Whatever the social forces leading to superficiality the effects on the constituent members of the field are profound. One of the basic assumptions on which all social fields rest is that:

The mutual confrontation of A and B attests to *their basic psychological similarity*.

Both are forced to assume that the other is like themselves in ways that distinguish them both from non-humans. This basic similarity does not rest simply on the perception of the contours, shapes, colours and textures of each other, but upon the recognition that both are subject to and behave according to similar psychological laws. In practically all kinds of behaviour—in laughing, loving, working, desiring, thinking, perceiving etc.—there are basic similarities between people that are open to the observations of others and underly our understanding of differences. Central in this perception of others is the awareness that they too can establish directive correlations with parts of the environment, i.e. they appear in the environment as action centers. The focal condition of their directive correlations (their goals or intentions) are revealed by such behaviours as change of direction at obstacles, cessation of activity on reaching a point, convergence of different means actions. Because behaviour is the key way in which these action centers manifest themselves 'it tends to engulf the total (Perceptual) field rather than be confined to its proper position as a local stimulus whose interpretation requires the additional data of a surrounding field.' (Heider, 1958, p. 54). In this perception of the activities of others it seems also that the manner of carrying out the act conveys, in a direct and open way, the experience of the acting person. His hesitancy, striving, euphoria etc. are normally manifest in his behaviour.

The perception of basic psychological similarity underlies the assumption, likely to be held by A and B, that anyone else similarly placed would see what they see, feel what they feel and do what they do. On the same grounds A can grant the possibility that B can provide him with relevant information about aspects of X which are revealed to B in his position but not to A. Disagreements between A and B about X are likely, because of assumed basic similarity, to create a desire for deeper knowledge unless some special individual difference can be imputed (e.g. colour-blindness or difference in interest). The strength of the psychological forces thay may be aroused by such disagreements is well illustrated by Asch's experimental studies. As the subjects of these experiments were forced to accept the reality of the challenge to their assumption of 'naive realism' they tended to question the assumption of psychological similarity; to suspect some defect in themselves. 'These circumstances fostered an oppressive sense of loneliness which increased in prominence as subjects contrasted their situation with the apparent assurance and solidity of the majority'. (Asch, 1956, P. 32).

It is the prevalence and the potency of this form of passive adaptation that leads Marcuse (1964) to characterize 'advanced industrial society' as *One Dimensional Society* and its typical citizen as *one dimensional man.* Like us, he starts from the point that 'the range of choice open to the individual is not the decisive factor in determining the degree of human freedom, but *what* can be chosen and what *is* chosen by the individual'. The latter is not restricted by suppression or repression but 'the distinguishing feature of advanced industrial society is its effective *suffocation* of those needs which demand liberation' (Marcuse, 1964, p. 7; my italics). In case he should be misread to imply that he is referring to the more trivial consequences of the mass media presenting an over-complexity of choice, Marcuse emphasizes that:

The pre-conditioning does not start with the mass production of radio and television and with the centralization of their control. The people enter this stage as pre-conditioned receptacles of long standing; *the decisive difference is in the flattening out of the contrast (or conflict) between the given and the possible, between the satisfied and the unsatisfied needs* (Marcuse, ibid., p. 8; my italics).

This is what we mean when speaking of increased superficiality—of increased indifference to what needs or demands are taken as the starting point for one's behavioural responses. This is not only an individual response to over-complexity. A business conglommerate can diversify its product lines so that it can become relatively indifferent to the fate of any particular one. In a society it encourages '*fractionation*'—an arbitrary tearing apart of the social fabric. Members are thrust aside or move aside, not because they constitute a viable social sub-system with goals in conflict with the larger system, but because as individuals they are nonconforming. They refuse to be indifferent to the roots of their individual behaviour and are outcast as alcoholics, perverts, beatniks or eggheads. This maladaptive mode can become such an unquestioned part of social life that integrated communities of schools and families can tear themselves apart on the length of boys hair, without even laughing while they do so. Insofar as the non-conformists are an identified source of social variance, then their exclusion seems to reduce the total amount of relevant variance in the environment.

Marcuse goes beyond this in one very significant respect. I argued only that, given a turbulent environment, this way is one of three ways of passively adapting to it. He argues that this mode of adaptation has become so deeply rooted, at least in the U.S.A., that that society can be characterized as a 'one dimensional society' and, further, that it means that 'liberation of inherent possibilities' no longer adequately expresses the historical alternatives' (Marcuse, p. 255), or, in his final sentence, the quote from Benjamin: 'It is only for the sake of those without hope that hope is given to us' (Marcuse, p. 257).

Marcuse might be right about the present but I will stay with my earlier theoretical position and maintain that, while at this level of analysis one can spell out the alternative future paths, it is necessary to consider the leading part if one wishes to see what paths are likely to be taken. His judgement is, of course, very relevant even if it specifies only one of the present conditions from which men in the advanced industrial societies choose their futures. On this particular point, we have the reinforcing evidence advanced by Angyal. Experience in clinical practice up to his death in 1960 led him to observe that while

> . . . the dimension of vicarious living (hysteria) can be safely described as the 'neurosis of our times' . . . . Recently, however, the compliant (conforming) pattern emphasized by Fromm, Riesman and others began to give way to the secondary type, the hysteria with negativistic defences. The 'rebellious hysteric' is already quite prominent both in therapists' offices and on the social scene. It is possible that he will become the dominant sociological type, the spokesman of the times' (Angyal, 1966, p. 154: our inserts).

He sees the phenomena of the beatnik and hippy as 'a protest against the levelling tendency of social conformation which threatens the extinction of spontaneous individuality' (Angyal, p. 154).

From my point of view this changing pattern of common neuroses suggests that the neurotics may, like artists, be reacting to emerging trends before their more stable fellows. Their basic sense of personal worthlessness may make them more dependent upon the fabric of cultural symbols and hence more sensitive to flaws and rents that are beginning to emerge. What is today's preoccupation with T-grouping and teamwork may be the response to the neurosis of yesterday. What is today's neurotic striving for individuality may well be tomorrow's goal (or confusion). If this is so, then despite the impression that Marcuse and Angyal have of the dominance of superficiality, and the resulting fractionation of the social field, the forces toward other choices may already be operative in the advanced industrial societies.

Segmentation

The second way of simplifying an over-complex environment is that of *segmentation,* or more literally separation of the parts. As a social system differentiates to cope better with complexity, it also increases the possibility of parts pursuing their ends without respect to the total system. This may not be as big a threat to the survival of the part as at first it seems. Given a multiplicity of specialized parts, there may be many different assemblies of parts that can serve the system goals (crudely, troops can load ships if the dockers strike, or an airlift can be laid on). Thus, temporary non-functioning of a part need not lead to its permanent destruction or replacement. For the part itself the path of segregation involves the risk of major errors but these may seem no worse than risking the devil of over-complexity. There is increasing uneasiness amongst social observers that the recent rapid advance in industrial societies has been leaving behind largish segments of their own societies (notably the poor), intensifying the pressures to disintegrate into smallcr, more culturally homogeneous entities (whether Negro, Welch, Bretons, the urban poor or the rural communities), and widening the gulf between the cultures of advanced and under-developed countries. As a response to over-complexity this may be adaptive in allowing sub-groups to draw closer to the roots of their own cultures, provided and insofar as there emerge other system relations which, while less binding, enable the enhanced self-control of the part to be guided by a knowledge of the state, capacities and goals of the total system. Such system relations are emerging in national planning etc. and, although they may be a step behind the tendencies to segmentation, there is no convincing evidence that this is other than temporary. The same cannot be said of the relation between the 'developed' and 'underdeveloped' economies.

The criteria for measuring the growth of this form of maladaptation seem simple enough and ready grist for the scientific mill, namely the degree of coherence and regular communication between the segments or parts of a social field. The difficulty lies in appreciating the relevance of segmentation. It has always been a characteristic of disturbed reactive environments and, while deplored in its extreme forms (e.g. Myrdal, 1941) it has rarely constituted a major obstacle to the evolution of adaptive social forms. I doubt whether this is any longer true in turbulent environments. Turbulent environments not only provide an immense challenge to the adaptive characteristic of man and his institutions but, at a guess, contain the seeds of a further environmental evolution to predominantly vortical social processes where there can be no question of the constituent members adapting and evolving but only a possibility of sheer survival. Vortical processes emerge when one part of a field moves very much faster than the contiguous parts. They emerge at the boundaries of these parts and, willy nilly, tear off and draw in fragments from the adjoining parts. Long before such vortical processes become dominant they exercise a retarding effect on overall social progress out of all proportion to that which would be predicted from the ordinary social resistances to change. It is 'drawing a long bow' to suggest that such phenomena as the U.S. race riots, the Weathermen bombings and Vietnam are essentially vortical processes at the interface of white-coloured communities, young and old, East and West. It is not so fanciful to suggest that the mal-adaptive mode of segmentation may be a more serious obstacle to active adaptation in the future than more visible modes of superficiality and dissociation.

Dissociation

This form of passive adaptation is essentially an individual response to complexity and is supported by what happens at a social level, rather than induced by socially organized forces. It occurs when individuals seek to reduce the complexity of choice in their daily lives by denying the relevance or utility of others as co-producers of the ends they seek to attain. The pay-off, the apparent adaptiveness of this mode, is some sort of positive function of the extent to which the individual lives his daily life, travelling, eating, working, entertaining in mass conditions. Little wonder that dissociation (anomie) has always been a charge laid by traditional societies at the gates of the cities. In turbulent environments it is not so much the mass character and anomie per se that encourages this mode of passive adaptation but the increasingly unpredictable nature of what might follow from even seemingly trivial involvements with others. It is quite possible that there are cultural differences. Certainly the British society used to seem remarkably more tolerant, less given to segmentative tactics than, for instance, the U.S.A. or Australia. On the other hand the British seem more likely to retreat to their suburban castles and dissociate, on the grounds of 'I don't want to know', while the Americans and Australians defend their superficiality with 'So what?', or 'I couldn't care less'.

Because this mode of maladaptation is essentially a personal mode its major effects on the integration of social fields are not direct ones. The major effects are manifested through changes in superficiality and segmentation.

Insofar as people withdraw from public commitments the roots of their behaviours are less open to the understanding of each other. They are compelled to judge each other in terms of the extrinsic characteristics of their behaviours. Their interrelationships become more superficial.

Dissociation means a reduction in the average man's sense of responsibility for co-ordinating and regulating his behaviour with respect to the potential coproducers of his desired ends. For each such fractional reduction there is a marked multiplier effect. Special and massive social regulatory institutions have to be created, de novo or by expansion of existing bodies, to carry responsibilities formerly implicit in the web of mutual support that constituted the social field. Each such move, however justified to maintain the integration of the social field, seems to yield but another form of segmentation and to reduce the chances for active adaptation.

Although, for reasons we have suggested, dissociation has not got the same social visibility as the other mechanisms Neuman (1954) sees it as being at least as important in modern society as superficiality. He points to the loss in power and intensity of the cultural canons (e.g. 'God' and 'conscience') which once defined a common world for joint action. The trends in criminal statistics, road accidents, abortions and personal liability suits certainly suggest that there are forces in the society that support Neumann's viewpoint.

Reviewing our notes on these three mechanisms, we can conclude:

(a) they are mutually facilitating defences, not mutually exclusive;
(b) they all tend to fragment the spatial and temporal connectedness of the larger social fields and focus further adaptive efforts on the localized here and now;
(c) they all tend to sap the energies that are available to and can be mobilized by the larger systems and otherwise to reduce their adaptiveness.

It is the operation of these processes that undermine the quality of life in a social field despite the relative affluence of the inhabitants.

Despite the strong cases that have been argued for superficiality and dissociation as major characteristics of the present, and despite the growing significance of segmentation, we do not think these necessarily define man's future. They are so important that any society should collect statistics on these processes as avidly as they collect meteorological data. However, none of the modern industrial nations is so obviously undermined by these processes that they lack the power to adapt in other ways.

Chapter 6

Active Adaptation—The Emergence of Ideal Seeking Systems

Two Basic Principles of Social Design
Currently Favourable Trends in the Leading Part
Matrix Organization
Strategic Planning
Summary

Men are not limited simply to adapting to the environment as given. Insofar as they understand the laws governing their environment they can modify the conditions producing their subsequent environments and hence radically change the definition of 'an adaptive response'.

Such possibilities are present in turbulent environments. There are some indications of a solution which might even have the same general significance for these turbulent environments as the emergence of strategy (or ultra-stable systems) has for clustered and disturbed reactive systems. Briefly, this is the emergence of values which have an over-riding significance for members of the field. Values have always arisen as the human response to persisting areas of relevant uncertainty. Because we have not been able to trace out the possible consequences of our actions as they are amplified and resonated through our extended social fields, we have sought to agree upon rules such as the Ten Commandments that will provide each of us with a guide and a ready calculus. Because we have been continually confronted with conflicting possibilities for goal pursuit, we have tended to identify hierarchies of valued ends. Typically these are not just goals or even the more important goals. They are ideals like health and happiness that, at best, one can approach stochastically. Less obvious values, but essentially of the same nature, are the acioms and symbols that lead us to be especially responsive to certain kinds of potentialities. Prejudice is a clear example of this latter kind of valuation; pride in rationality or in personal autonomy are cases where the evaluation concerns potentialities within oneself.

It is essential to bear in mind that values are neither strategies nor tactics and cannot be reduced to them. As Lewin has pointed out, they have the conceptual character of 'power fields' and act as guides to behaviour.

> Values influence behavior but have not the character of a goal (that is, of a force field): For example, the individual does not try to 'reach' the value of fairness but fairness is 'guiding' his behavior. It is probably correct to say that values determine which types of activity have a positive and which have a negative valence for an individual in a given situation. In other words, values are not force fields but they 'induce' force fields.

Insofar as values do emerge, the character of the richly joined turbulent fields changes in a most striking fashion. For large classes of events their relevance no longer has to be sought in an intricate mesh of diverging causal strands, but is given directly and in almost binary form by references to the ethical code. So clear and direct is this form of reference that men have typically failed to distinguish between the value and its various physical and social symbolizations (Goldschmidt, 1959, p. 76).

> One of the insistent features of man's symbolization is that symbol and reality become so intertwined and unified in the thought of the people that they are not easily separated: one cannot readily determine what is symbol and what reality. Ask a Plains Indian what makes a good man, and he may tell you bravery; he may tell you 'counting coup'. We have treated values as the desirable qualities in human character as defined by the culture, but these inevitably are vaguely expressed qualities; abstract, intellectual, and rather remote from the activities of daily life. The symbolic elements are titles, memberships, rituals of deference, forms of dress and adornment, possessions: the concrete attributes which have more dramatic effect upon the personnel of the culture.

By this transformation there is created a field which is no longer richly joined and turbulent but is simplified and relatively static. Men and their organizations can expect to adapt successfully to this transformed field.

In suggesting that values offer a way of coping with our emerging turbulent environment, we have only opened up the problem, but at least it directs attention to a set of subordinate questions. 'What values?', 'How do these values enter into and shape the life of the individual?' and 'How do these values enter into and shape the organizational structures that men create?' The first of these questions is the most difficult; the others only somewhat less so.

The difficulty with the first question is quite simply that we have done so little to establish a *'science of morality'* (Chein, 1947). What we do know about values is that they take a tremendously long time to emerge. The salience of a particular value may change rapidly both for a community or an individual, but a new value can be distilled only from generations of experience. This unselfconscious process of value formation is too slow to meet present pressing requirements. It seems necessary for social scientists to exert their efforts to speed up the distillation process, although at the moment the most concrete proposals we have for identifying ideal goals are those of Churchman and Ackoff (1949, pp. 327-39). Short of this, something can be done by searching from amongst existing values for some that may be appropriate. This can only be an *ad hoc* solution fraught with dangers. If it is necessary to beat a partial retreat from the overwhelming uncertainties of a turbulent field, then it is nevertheless crucial that *the substitute symbolical field represents in its key symbols—the values—the main causal strands of the real world.* The existing Western values may not convey enough of the new realities and, in any case, we still have to develop methods of analysis that will identify the referents of values. On general grounds we may well query whether existing values provide an adequate pool from which to select. The processes of social evaluation have too frequently proceeded from an initial simplifying dichotomy of God or the Devil. This sort of distinction usefully goes beyond the notion of good or evil because it allows that what seems to be good is evil and vice versa. The simplification to *external* competing influences has, however, meant little development of values as guides in the areas where organized social life and group life are both critically involved—in the areas that we tend to label as charismatic, mob behaviour, fads and fashions or otherwise as irrational group behaviour. These sorts of blind ground-movements would seem to be salient in the turbulent fields, but inexplicable in terms of external influences such as 'the workings of God' or 'the machinations of the Devil' (see Pareto's residues, 1924).

If these questions about values each had to have its own

separate solution, we might well doubt whether men could cope with them in the next generation and then find ourselves writing some very pessimistic scenarios for the 1990s. In my view this is not the case—the three problems seem to be soluble by a single strategy.

Two Basic Principles of Social Design

This strategy is based on the notion that it is in the design of their social organization that men can make the biggest impact upon those environmental forces that mould their values (that make some ends more attractive, some assumptions about oneself and one's world more viable): further, it assumes that if these changes are made in the leading part, the socio-technical organizations, the effects will be more likely to spread more quickly than if made elsewhere. I realize this is contrary to the Billy Graham strategy of going straight to the hearts of men and that it is contrary to Jesuitical-psychoanalytical notions of going to the cradle or school. I am suggesting that *adults be the educators and that they educate themselves* in the process of realizing their chosen organizational designs. This confronts us with the question of what values, and we are suggesting that the first decisions about values for the future control of our turbulent environments are the decisions that go into choosing our basic organizational designs. If we can spell out the possible choices in design we can see what alternative values are involved and perhaps hazard a guess at which values will be pursued by western societies.

As this spelling out has to be stretched out and may be a bit tedious, I will state my conclusions first. A choice in basic organizational design is inevitable so there is no question but that men will make them (even if they are not conscious of doing so); *the choice is between whether a population seeks to enhance its chances of survival by strengthening and elaborating special social mechanisms of control or by increasing the adaptiveness of its individual members.* The latter is a feasible strategy in a turbulent environment and one to which western societies seem culturally biased.

Choice seems unavoidable. What makes it unavoidable is what might clumsily be called a *design principle.* In designing an adaptive self-regulating system, one has to have built in redundancy or else settle for a system with a fixed repertoire of responses that are adaptive only to a finite, strictly identified set of environmental conditions. This is an important property of any system, as an arithmetical increase in redundancy tends to produce a logarithmic increase in reliability. Redundancy may be achieved by having *redundant parts,* but then there must be special control mechanisms (specialized parts) that determine which parts are active or redundant for any particular adaptive response. If the control is to reliable it must also have redundant parts and the question of a further control emerges. In this type of system, reliability is bought at the cost of providing or maintaining the redundant parts, hence the tendency is toward continual reduction of the functions, and hence cost, of the individual part. The social system of an ant colony relies more upon this principle than does a human system, and a computer more so than an ant colony. The alternative principle is to increase the *redundancy of functions* of the individual parts. This does not entail a pressure toward higher and higher orders of special control mechanisms, but it does entail effective mechanisms within the part for setting and re-setting its functions—for human beings, shared values are the most significant of these self-regulating devices. Installing these values of course increases the costs of the parts. The human body is the classic example of this type of system although it is becoming more certain that the brain operates by means of overlapping assemblies based on similar sharing of multi-skilled parts.

Whatever wisdom one attributes to biological evolution, the fact is that in the design of social organization, we have a genuine choice between these design principles. When the cost of the parts is low (in our context, the cost of individual life), the principle of redundant parts is attractive. The modern Western societies are currently raising the value of individual life, but a change in reproductive rates and investment rates could reverse this. There is, however, a more general principle that favours the western ideal. The total error in a system can be represented as equal to the square root of the sum of the squares of all the component errors. It follows that a reduction in the error of *all* the components produces a greater increase in reliability than does an equal amount of reduction that is confined to some of them (e.g. to the special control parts). We are certainly not suggesting that this principle, per se, has been or is even now a conscious part of western ideologies. Some sense of it does, however, seem to have reinforced our prejudice toward democratic forms of organization and our prejudice against elitism.

Two further factors operate in the same direction. When the sources of error are not independent, i.e. they are correlated, then the tactic of overall reduction in error is even more advantageous. In human systems, communication is a potent factor and hence the advantages are considerable. When, in fact, the alternative design principle of redundant parts is adapted, there are strong reasons for reducing the correlation of parts (e.g. anti-unionist practices, the censorship of the mass media in totalitarian societies and the management of concentration camps). The second factor also happens to be our basic concern—environmental complexity. The second design principle allows for a much greater range of adaptive responses than does a redundancy of parts. Although its tolerance for error in any particular response is less, this is a condition for greater learning.

Whatever the advantage to the individual of organizational designs based upon redundancy of functions and despite the sum of the advantages we have mentioned, it is by no means certain that this gives survival advantages to the total international system. Whether it is or not, we will be better able to judge by the end of the next thirty years when, with the industrialization of Asia, there will have been a more equal test of the alternatives. In any case, it seems much more likely that the western societies will seek solutions in this direction, to the point of non-survival, than that they will evolve to some sort of Orwellian 1984. A judgement of this kind does presuppose what I have not yet discussed—the character and likely development of the leading parts of the system.

Currently Favourable Trends in the Leading Part

Systems Management

Certain current developments in the area of technology/production give us reason to hope that effective 'democratic' solutions will be found before the passive adaptive modes force us toward 'totalitarian solutions'. These are the rapid emergence of, in the U.S., what has been termed 'systems management', and the programmes being pursued in the U.K. and Norway by trade union leaders and management to develop (with social scientists) effective ways of involving individuals in the control of their working organizations (Emery, 1967). Systems management is a radical change from our traditional patterns of organizations and much wider in its concerns and application than the much advertised cost-effectiveness studies of weapon systems. Its characteristics clearly relate it to the general problem of environmental transformation that we have been describing:

1. A more open and deliberate attention to the selection of ends toward which planned action is directed, and an effort to improve planning by sharpening the definition of ends;
2. A more systematic advance comparison of means by criteria derived from the ends selected;
3. A more candid and effective assessment of results, usually including a system of keeping track of progress toward interim goals. Along with this goes a 'market-like' sensitivity to changing values and evolving ends;
4. An effort, often intellectually strenuous, to mobilize science and other specialized knowledge into a flexible framework of information and decision so that specific responsibilities can be assigned to the points of greatest competence;
5. An emphasis on information, prediction and persuasion, rather than on coercive or authoritarian power, as the main agents of co-ordinating the separate elements of an effort;
6. An increased capability of predicting the combined effect of several lines of simultaneous action on one another; this can modify policy so as to reduce unwanted consequences or it can generate other *lines* of action to correct or compensate for such predicted consequences (Way, 1967, p. 95).

As a response to the complexity of large scale organizations:

the new style can deal with that by distributing to a larger and larger proportion of the population responsibility for the decisions that shape the future. It can also inculcate a common style of action among business managers, government officials, and university professors; already, more and more people are circulating freely through all three of these formerly walled-off worlds. By mobilizing specialized and value-free science to work on practical problems, the new pattern can help restore the community of scientists and scholars and build an organized link between science and value (Way, 1967, p. 95).

This development has not taken place without its confusions. These confusions have stemmed largely from false assumptions about computers as artificial intelligences and about the omniscience of experts. Given these assumptions, systems management can be conceived of as a great strengthening of the totalitarian design. It has taken time to realize that:

(a) Decision making and judgement cannot be reduced to the narrow band of formal logical structures to which computers are restricted (Cowan, 1965; Dreyfus, 1965).
(b) 'Optimization techniques can take into account only those uncertainties concerning the future that can be identified beforehand. Through optimization, furthermore, we can develop a control unit or monitor to be *added to the system* to deal with these predictable uncertainties—but we cannot provide a control unit that is *built into the system,* leading to increased self-control of the units already in the system' (Ackoff, 1966).
(c) The rationality of a social system is not a property of an isolated part (however expert); it is a property of the system of which the experts are only a part, occupying a position in relation to all the other parts. The design of inquiring sub-systems has become one of the very pressing problems because 'where centralized planning begins to narrow the ability of individuals to express themselves in certain traditional ways, then the system has become less effective and the system scientist should translate the lack of freedom in the system into a deterioration of the system effectiveness' (Churchman, 1966).

Matrix Organization

Systems management and the U.K.-Norwegian experiments are still very small developments, and it may seem unwise to read too much into them. I have felt more confident in this interpretation because it has been possible to identify some features of the general organizational response that would be adaptive in turbulent environments. What stands out from our own experience (not least from our attempts to devise a more appropriate organization for our own peculiar social situation) is that the characteristics of the turbulent field require some overall form of organization quite different to the hierarchically structured forms to which we are accustomed. Whereas the Type 3 environments require one or other form of accommodation between like but competitive organizations (whose fates are to a degree correlated), the turbulent environments require some relationship between dissimilar organizations whose fates are basically positively correlated: that is, relationships that will maximize co-operation while still recognizing that no one organization could take over the role of the other. For obvious reasons I am inclined to speak of this type of relationship as an *organizational matrix:* it delimits the shape of things within the field it covers, but at the same time, because it delimits, it enables some definable shape to be achieved. While one aspect of the matrix provides a conference within which the ground rules can be evolved, another independent but related aspect must provide for the broader social sanctioning. Insofar as the sanctioning processes can be concretized in an institutional form, it should be possible for the conferences to have the degree of secrecy and protection that is required if the component organizations are to retain an effective degree of autonomy and engage in effective joint search for the ground rules. It is possible to foresee that within the domain covered by such a matrix there would need to be further sanctioning processes to control the diffusion of values throughout the member organizations. This appears to be one of the functions exercised by professional bodies.

It should be noted that, in referring to the matrix type of organization as one possible way of coping with turbulent fields, we are not suggesting that the higher level sanctioning can be done only by State controlled bodies, nor are we suggesting that the functioning of these matrices would eliminate the need for other measures to achieve stability. Matrix organizations, even if successful, would only help to transform turbulent environments into the kinds of environments that we have discussed as 'clustered' and 'disturbed-reactive'. Within the environments thus created, an organization could hope to achieve stability through its strategies and tactics. However, the transformed environments would not be quite identical. Thus the strategic objective in these transformed environments can no longer be stated in terms of optimal location (as in Type 2) or capabilities (as in Type 3). The strategic objective has to be formulated in terms of institutionalization. As Selznick states in his analysis of the leadership of modern American corporations: 'the default of leadership shows itself in an acute form when *organizational* achievement or survival is confounded with *institutional* success' (1957, p. 27): '. . . the executive becomes a statesman as he makes the transition from administrative management to institutional leadership' (1957, p. 154). This transition will probably be rendered easier as the current attempts to redefine property rights clarify the relations between the technologically productive area and the total social system. Private property rights are being increasingly treated as simply rights of privileged access to resources that still remain the resources of the total society. To that extent, the social values concerning the protection and development of those resources become an intrinsic part of the framework of management objectives and a basis for matrix organization.

Strategic Planning

The processes of strategic planning are also modified. Insofar as institutionalization becomes a prerequisite for stability, then the setting of subordinate goals will necessitate a bias toward those goals that are in character with the organization and a selection of the goal paths that offer a maximum convergence of the interests of other parties. Hirschman and Lindblom (1962) have spelt out in some detail the characteristics of policy-making under these conditions of environmental complexity, uncertainty and value conflict. Detailed studies of the decision processes in large scale systems lead to the view that these processes are most effective when they allow for the co-ordination that arises from the mutual adjustment of the values and interests of the participants even though these social processes may not be consciously directed at an explicit common goal, and decision processes are characterized by *disjointed incrementalism.*

Summary

What we have been predicting is the emergence of a process, not a particular event. We think that the outlines of the process can already be detected and that it is a process which could evolve both the values and organizational structures which can transform our present social environment. If one wishes to predict in more detail, it is necessary to consider the more closely the technological/productive area, its leading part, the informational technology, and those characteristics of the other social areas that will affect the diffusion of change throughout the system. If one wishes to qualify these predictions, it would be necessary to consider the wider international setting. Time prevents further analysis at this level of detail. We can only list the matters we would have wished to deal with.

In the technological/productive area, the significant changes include:

(a) growth in G.N.P.; (b) growth in productivity; (c) growth in range of what can be produced; (d) increasing indirectness of human contribution to the productive process.

Among the social and human effects that need to be considered are:

(a) the changes in the salience of human affects as distress becomes less dominant. Cultural differences should be considerable; (b) the shift in balance between the portion of life given over to work and leisure; (c) the shift in balance between the Man-Nature, Man-Man relations.

In the field of information technology the significant changes are:

(a) the shift in balance of costs between communication and transportation; (b) computerization of an increasing portion of object-object relations and man-object relations where man can appropriately be considered as an object (e.g. allocating a man to an aircraft seat): this makes possible a shift in salience; (c) the shift in balance of distal and proximal communications.

The range of social and psychological effects may be no less extensive than what one would expect from a major mutation of the species. Of particular concern are the effects on man's perception of himself and his world. As Arendt and Kuhn have argued, these types of changes are fundamental in the evolution of society and of science. It is assumptions about these things that tend to determine the way men use and develop their technological apparatus.

Because information technology is the leading part in the technological/productive system, we can expect it to have a major formative influence upon work and learning for work. We would certainly expect the nature of work and learning to change and it is possible that the radical changes in information technology are producing radical changes in these fields.

The boundary conditions of the modern industrial societies are not likely to remain constant throughout the next thirty years. Two main sets of conditions have attracted attention: (a) qualitative and quantitative changes in the population inputs to the world society; (b) qualitative and quantitative changes in the other resources available to the world society.

These variables are not independent of social action and hence cannot be predicted from their previous trend lines. The modern industrial societies are such a leading part that their own actions can affect these variables. This creates for them a range of relevant choices. They are, however, still only a part of the world society. The choices they make will be moulded by the relationships they develop with the others, particularly as their individual fates are becoming more closely integrated, and their contacts increasing. These very conditions may reveal deep cultural fissures that were irrelevant in the earlier imperialist phase but are now becoming critical.

It should be possible to explore the effects these types of changes could have upon future development (Emery and Wilson, 1971).

Throughout, there has been no attempt to identify the particular contributions that social science should make. We have assumed that the first task was to identify ways in which our anticipations could be improved, secondly to venture a few of the broadest guesses. It would be a further task to see what specific social science developments would best help meet the anticipated problems and possibilities.

Part two

Aspects of the transition to post-industrialism

E L Trist

Re-evaluating the Role of Science[1]

Positive or Negative Science Policies: A Strategic Choice
New Concepts of Science
New Concepts of Policy

Positive or Negative Science Policies: A Strategic Choice

Since the Scientific Revolution in England in the 17th century the cultural process we know as science has been proceeding at an exponential rate. Its effects on technology have transformed not only Western societies but the world environment. Having altered man's conception of the universe and of himself it has altered also his chances of survival. For it has enabled him to produce the weapons to destroy himself, the medicine to jeopardize his food supply by overpopulation, and industrial products irreversibly to despoil his habitat. By that same token it has given him the means of passing beyond the state of bondage to which he has been historically accustomed to a society where the quality of life could be of a different and higher order.

Few would care to predict which of these destinies will be ours or what dangerous and unstable mixtures of the two we may have to endure before (if ever) a saner balance is struck. Science, through which the 19th century so confidently thought we could control the world, is seen in the latter part of the present century as the means through which we are making the world uncontrollable. As Sir Geoffrey Vickers (1968) has put it, 'we have reached the end of free fall'. The world which our scientific culture has been making is no longer

[1] This chapter and chapter 8 draw on material originally presented in the Background Paper for the Round Table on the 'Social Aspects of Science Policy', March 1969, University of Toronto, Harry M. Cassidy Memorial Research Fund.

auto-regulative. This scientific culture has begun to disturb a number of the balances in the social and bio-physical ecology on which as a species we have depended. The forces unleashed have become too powerful to be contained simply by the natural interplay of other forces. To understand why this is so requires an appreciation of what has become the salient characteristic of the contemporary environment, namely, that it is taking on the quality of a turbulent field (chapter 4).

The root question concerning the social aspects of science policy is this: can science, which has been the *sine qua non* among factors leading us towards a more unregulable world, be the *sine qua non* also in leading us towards a more regulable one? A negative answer would have as a consequence advocacy of a negative science policy—a withdrawal of resources from science and its dis-establishment as a core value in the culture of our society. A positive answer would have the consequence of advocating a positive science policy—the investment of increased resources in its already extensive domain and its even firmer establishment as a core value.

Whatever else it may claim to be the present is the age which has brought science into politics. Positive science policies are becoming more common and more comprehensive in ever more countries. The escalation of government scientific expenditure during the '60s has been greater than in other fields. Yet there are signs in several countries of a turning away from science in the present student generation. A number of places, for example, in science and technology have not in recent years been taken up in British universities. The abandonment of what is perceived as the scientific ethos is a prominent feature of many of the counter-cultures in the United States. In the 'search for relevance' (Axelrod *et al.,* 1969) which is going on among members of university faculties as well as students science is frequently attacked from the left as ultimately responsible for producing a depersonalized technological society while from the right it is attacked by those who want to arrest change and fear science as the agent of change. We cannot take it for granted that our societies will continue to support positive scientific policies.

One of the consequences of government supplying the bulk of R & D funds is that the ordinary citizen will have an increasing say, however indirectly, in the affairs of science. For, in democratic countries at least, the continuation of a government depends on the support of the majority of the citizens. Their intuitive appraisals of scientific, as of more familiar, issues must be taken note of. This is an aspect of science having become political which has been less discussed than that of some constraining of scientific choice as regards what the scientist may himself do. It is an innovation, however inadvertent, which may constitute a safeguard. If on the one hand it makes it easier for science to be regarded as part of 'the establishment', on the other, it makes it more likely that the uses of science will, ultimately, be subject to democratic control than if the bulk of the funds were privately provided.

Western societies (and in the longer run others) will not continue to support positive science policies unless the world view of science can be shown to be reconcilable with human values. Ordinary Western man—far from ill-educated now and likely to become far better educated—requires evidence that science can at bedrock be trusted to work in the human interest. Yet the increased complexities of the contemporary environment cannot be understood—far less their instabilities controlled or their potentiality for beneficient change realized—without the assistance of a science developed beyond its present capabilities. The costs of abandoning a positive science policy would be penal. A case, however, for continuing a positive policy can be made, in the writer's view, only on the basis of the social aspects. What follows will outline the case.

New Concepts of Science

Three changes have occurred which have made science more 'human' than it seemed several decades ago. These are the abandonment of the belief in total explanation, the abandonment of reductionism and the appearance of an integrative, in addition to an analytic, strategy.

The scientific world view which prevailed in the 19th century, and which still haunts the popular image of science, was not reconcilable with human values. For a model based on mechanism, atomism and determinism scarcely depicted a world which men could live in when other possible worlds were ruled out. The coming of the relativity and uncertainty principles upset this view in the physical sciences. The more sophisticated concepts that have followed have removed omniscience from science, putting a limit on the realm of scientific explanation. Michael Polanyi (1967) has summed up the paradoxical result as follows:

> The current situation in the philosphy of science is a strange one. The movement of logical positivism, which aimed at a strict definition of validity and meaning, reached the heights of its claims and prestige about 20 years ago. Since then it has become clearer year by year that this aim was unattainable. And since (to my knowledge) no alternative has been offered to the desired strict criteria of scientific truth, we have no accepted theory of scientific knowledge today.
>
> Take Ernest Nagel's widely accepted account of science. He writes that we do not know whether the premises assumed in the explanation of the sciences are true; and that were the requirement that these premises must be known to be true adopted, most of the widely accepted explanation in current science would have to be rejected as unsatisfactory. In effect, Nagel implies that we must save our belief in the truth of scientific explanations by refraining from asking what they are based upon. Scientific truth is defined, then, as that which scientists affirm and believe to be true.
>
> Yet this lack of philosophic justification has not damaged the public authority of science, but rather increased it. Modern philosophers have excused this unaccountable belief in science, by declaring that the claims of science are only tentative.

This means that the scientist can no longer lay claim to the whole truth. There are other forms of understanding. This situation had to exist before scientists, professionals, administrators and politicians could collaborate in relations of mutal respect. It is a necessary condition for a positive science policy, which depends on bringing together these four groups, called by Price (1965) 'the four estates', in a context which can lead to the taking of *informed social action.*

Next, the advent of open systems and information theory in biology upset the principle of reductionism. Other options than closed system models are now available to explain the negative entropy of the 'living'. Moreover, these advances in biology are assisting the social sciences in finding their own conceptual identity in approaching more appropriately the psycho-social worlds created by men in their societies. Concepts such as 'appreciation' suggested by Vickers (1965, 1968) no longer require apology. All these steps are apprehended as being within science, which seems more self-consistent if less unified. Science as a method, as an inquiring system, has liberated itself from the domination of the physical sciences, indeed from 'scientism' (Kaplan, 1964; Churchman, 1967). As Vickers has said of the distinctive domain of the social sciences:

> In the human species the responsiveness (of the living organism) has become the basis for a further development so far-reaching that it needs to be distinguished as a third stage, because it introduces not only a new means of mediating change but even a new dimension in which change can be mediated. This new dimension is the conceptual system whereby humans represent, interpret, value, and increasingly create the world in which they effectively live. The new mediator is human communication, notably dialogue and the internal procedures which have developed with its use. The conceptual system thus developed is a psycho-social artifact, of which the conceptual world created by science, with its attendant procedures, is the most stable, coherent, and explicit example. But business, politics, and other human activities have their own partly autonomous systems; and each individual from birth to death is to be found, in self-directing and self-limiting development, an individual system as unique as his genetic code but containing initially far more possibilities than can be realized. These developments, individual and social, know springs and forms of change which have no counterpart in the purely responsible organization of other creatures. I will label the new mediator of change 'appreciation'.

Science, it would seem, can come to homo sapiens without dis-establishing him from his human estate (Levi-Strauss, 1960). The scientist need no longer be feared as advocating the wrong 'design principle' as Emery has called it. Emery's exposition of the difference between the principles of the 'redundancy of parts' (the inanimate model) and the 'redundancy of functions' (the animate model) is given in chapter 6. Only the second can avoid an Orwellian world.

The appearance of an integrative strategy has shown that science has become able to cope with the reality of wholeness as well as that of elements. So long as it seemed to insist that only elements were real it did violence to a 'truth' intuitively grasped in human experience. The integrative strategy has emerged in terms of the 'systems' concept.

Russell Ackoff (1959) notes:

> In the last two decades we have witnessed the emergence of the 'system' as a key concept in scientific research. Systems, of course, have been studied for centuries, but something new has been added. Until recently scientists and engineers tended to treat systems as complexes whose output could be expressed as a simple function of the outputs of the component parts. As a consequence, systems were designed from the inside out. Increasingly

researchers have come to deal with systems whose output cannot be expressed as a simple function of component outputs and it has become more productive to treat them holistically and to design them from the outside in.

The systems concept has had a double origin, for it arose in 'systems engineering' as well as in theoretical biology. The sharing of a common concept has enabled pure and applied science to merge their activities in a way not previously possible. There have been two related effects:

(a) Disciplines of a new kind have arisen, such as operational research (more broadly called the management sciences) which are becoming linked to the social sciences (Lawrence, ed., 1966).
(b) These disciplines deal directly with technological and social systems in the complexity in which they exist.

This is a capability which science did not have at an earlier period. It has greatly increased its usefulness to those concerned with the management of human affairs just as the other advances mentioned have reduced the likelihood of its leading them into error.

New Concepts of Policy

If science has been changing so has policy. While the former has become more policy-aware the latter has become more science-aware. They have become 'directively correlated' in response to the increased uncertainties and inter-dependencies of the contemporary environment. These have had two effects on policy-making which have made it seek to become more 'science-based':

(a) The greater uncertainty requires more future-orientation;
(b) The greater interdependence requires more comprehensiveness.

When the change-rate was slower, policy could be largely corrective, acting after the event. With a faster change-rate, it has had to become more anticipatory, acting before the event. This relates it to planning. The task of government now extends from regulating the present to creating the enabling conditions for the future. This entails deciding how resources are to be committed without foreclosing too many of the options (Friend & Jessop, 1969), so that somehow this future may take place in one of the more desirable of its alternative forms. Such a task cannot be carried out without an extensive information base which can only be brought into being and maintained through the use of a wide range of sciences. Moreover, this task continuously challenges these sciences to develop new concepts and methods.

In the economic and social fields the first requirement is a more informed picture of the present, the state of which becomes more unknown the faster and more uneven the change-rate. Disaggregated as well as aggregated statistics and indicators are needed for short-run projections, the identification of high risk areas and the separation of the least from the most changing parts of the society. Beyond this, techniques have to be developed for detecting and interpreting emergent social processes and for constructing models of alternative futures (chapter 3).

When the sub-systems of society were less interdependent, policies could be more discrete and separate agencies could administer their own programmes with minimum reference to each other. The greater degree of interdependence has changed this situation. Diffuse problems now arise affecting several sections or indeed the whole of a society and these problems tend themselves to be interconnected. Examples would be poverty, obsolescence, urban decay, pollution, overpopulation, regional disparity, water and other natural resource management, inter-generational conflict. The complex perceptions and beliefs which arise about such interconnected areas (rapidly diffused by the media) Michael Chevalier (1967) has called 'meta-problems'. They are less accurately related to the source problems than when these were more bounded and linked to the direct experience of particular groups. Diagnosis is correspondingly more difficult. The causes and boundaries of these problems cannot be established without research.

The implications for policy have been stated, with reference to the United States, by Lawrence Frank (1967):

> The Federal Government now provides a wide range of professional and technical assistance, with many direct subsidies and special tax allowances and concessions to business, finance, industry, transportation, and communication–indeed, to the whole range of free enterprise. This

assistance to private business has been explained and justified as promoting prosperity and advancing the national welfare. But assistance and services to individuals and families have been strongly resisted and only reluctantly provided since there is no adequate rationalization for such extensions of government activities. The need for a political theory for this emerging 'Service State' is, therefore, especially urgent.

The Service State, not to be confused with the Welfare State with its aura of charity and philanthropy, is oriented to the enhanced 'well-being' of everyone, as Halbert Dunn has expressed it. It marks the acceptance of human conservation as the basic democratic task; each year sees the enlargement and extension of services furnished directly or financed by the Federal Government and reinforced by state and local agencies. These services embrace medical and health care, improved housing and urban rehabilitation, educational facilities and programs from early childhood into adult years, plus the improved care and support of the indigent, the handicapped, the impaired, and all others incapable of fending for themselves in our money economy.

Each addition and enlargement is made as a separate program with no coherent and systematic commitment, no political theory to justify and rationalize these enlarged government activities, and no statement of policy for their extension and administration. We are improvising and operating by a series of piece-meal programs.

This implies the need for an over-all, comprehensive policy that will assert the criteria for choices and decisions. With a clear statement of policy, those who make social decisions can be guided, as if by 'an unseen hand' when exercising their autonomy to integrate their efforts by collaborating with others who are reponsive to these same criteria. Without a statement of basic criteria for national policies, the various specialized programs and the separately located authority of governments and private agencies will continue to plan and execute their separate and often irreconcilable programs.

This orientation makes it apparent that we are moving towards another type of society than that to which we have been accustomed. This is sometimes referred to as the new service society, the society of the second industrial revolution or the post-industrial society. There is no guarantee of our safe arrival. Not only are the interdependencies greater—they are differently structured. To clarify value-confusion over welfare requires an understanding of its changing relation to development (chapter 10). Research by the social or whatever group of sciences is most appropriate is needed on all such problems. The changes in the policy field demand a new mobilization of the sciences. This could not be effected had not changes in science taken place which reconcile it with human values (c.f. Bronowski, 1961).

The Establishment of Problem-oriented Research Domains

The Character of Domain-Based Research
The Mapping of Domains
A Comparison of Domain Selection in Two Countries

The Character of Domain-Based Research

The changed relationship of science and policy and the changes that have been taking place in each have led to the emergence of a new type of scientific activity. This tends to be confused with more familiar types. Traditionally, the spectrum of scientific activities has been thought of as including fundamental research, applied research and development work. The economists have recently added another term, innovation. This refers to the additional activities that must be undertaken before the benefits of R & D can be realized in goods and services effective in the market place. The concept of innovation is also applicable to the non-market sector. Yet, if the ends of the spectrum are now clearer something has become blurred in the middle. For some time the term 'problem-oriented research' has been struggling into existence not knowing whether it should be subsumed under applied research or represent something different (de Bie, 1970). The thesis is advanced that it comprises a distinct category whose recognition has central importance for science policy and its social aspects.

If fundamental research is discipline-based, problem-oriented research may be said to be *domain-based*[1]. Domain-based inquiry links a group of sciences to a major sector of social concern. The problems are generic rather than specific. They

[1] For this use of 'domain' see McWhinney (1968).

give rise to meta-problems. They require on-going endeavour leading to cumulations of findings rather than 'solutions'. These findings contribute simultaneously to the advancement of knowledge and to human betterment. The development of a domain is jointly determined by the social and scientific interests concerned. From the policy standpoint such a domain has the characteristics of future-orientation and comprehensiveness. On the scientific side it involves the integrative strategy. Disciplines across the entire range of the physical, biological and social sciences tend to be drawn in. Their weighting and salience vary enormously between domains, which have very different centres and may evolve very different configurations.

Scientists, professionals, administrators and political representatives all become involved. The texture of their relationships differs from what it is in fundamental research, where scientific interest predominates, or applied research, where user-interest predominates. The relations of the different actors in a problem-oriented domain is that of collaboration. Bound together by common commitment to an overriding purpose they have to recognize the complementarity of their contributions and respect each other's authority. Domain-based research is a non-sovereign Type 4 process (chapter 14) based on the pluralistic surrender of power (chapter 16).

Domain-based problem-oriented research has experienced difficulties not only in securing recognition as a distinct activity but in finding appropriate organizational settings. Novel problems of decision-making and mutual responsibility are posed. These are not well understood and institutions which will allow the necessary experience to be gained to a large extent await development. This is scarcely surprising since domain-based research represents the confluence of emergent trends in both science and policy. Price (1965, op. cit.) gives the following account of the struggle of oceanography to find a place in the Federal scientific system of the United States:

Oceanography was the first large-scale federal scientific program. It began when Thomas Jefferson founded the Coast Survey in 1807 and employed a Swiss scientist, Ferdinand R. Hassler, to bring scientific instruments from Europe and begin the job of charting the seas for the guidance of navigators. Oceanography is a field of basic and applied science in which a great many federal departments and agencies have long been involved. But, for purposes of my story, the contemporary oecanographic program began in 1956, when a group of government oceanographers decided that their activities needed to be greatly built up. Indeed, the part of the story that I propose to tell begins in March 1961 when President Kennedy included an expanded oceanographic program in his first budget, and it ends twenty months later when he pocket-vetoed the Oceanography Act of 1962.

An idea of the scope of the expanded program is given by the various reports and testimony presented to Congressional committees. At least fourteen operating agencies were concerned, as well as the staff agencies in the Executive Office of the President.

The Navy, which had already revolutionized its strategic doctrine by developing the Polaris submarine and missile system, wanted more knowledge of currents and other ocean phenomena, both to increase its offensive capabilities and to defend against enemy submarines and their missiles.

The Geological Survey had its eye on the offshore oil on the continental shelf, and the Bureau of Mines on the promise of vast mineral resources in the ocean depths.

The Bureau of Commercial Fisheries was intrigued by the possibilities of increased protein supplies and even new kinds of food for an overpopulated planet. The Bureau of Sport Fisheries and Wildlife hoped to develop new recreational opportunities.

Medical researchers talked with excitement about the search in the oceans for new biological compounds that might give clues to the biochemistry of sanity and insanity—might even provide a clue for cancer. The Public Health Service, though it soft-pedaled such speculation, was concerned about pollution of our seafoods and our beaches and harbours by sewage and chemical wastes, as was the Atomic Energy Commission about the disposal of nuclear wastes.

Several agencies were interested in oceanographic research because of their roles in aiding navigation. The Coast and Geodetic Survey has the job of mapping the shores and currents; the Weather Bureau makes forecasts; the Corps of Engineers maintains harbours; and the Coast Guard keeps the sea lanes clear of dangers to shipping.

Finally, there were the established research and development programs. The Maritime Administration carries on studies to adjust the design of ships to oceanic conditions; the Smithsonian Institution conducts basic research; the National Science Foundation and other agencies make grants to universities and other institutions for a wide variety of investigation relating to the oceans.

The advocates of a comprehensive federal program knew that they were dealing not merely with a field of science, but a major problem in government organization. As Harrison Brown, chairman of the Committee on Oceanography of the National Academy of Science, told a Senate committee in 1960, the decision to be made on the organization of the oceanography program 'far transcends oceanography itself'. He noted that the undertaking, because of the way in which it cut across the programs of many operating agencies, typified the 'problem of decision making, concerning science and technology in Government'.

The Mapping of Domains

No systematic attempt has yet been made to describe problem-oriented research in terms of the domain concept or to relate such domains to the discipline-based fields of fundamental research or the user-prescribed missions of applied research. Obviously, one type of work can give rise to the others. If the overall scientific enterprise were to be mapped in domain terms, some fundamental, as well as applied, work would be included under domain headings. The Table below developed from an earlier Tavistock analysis (1964) is a trial exercise in this direction. It is based on a sector type concept as the most familiar, though not necessarily the best, and certainly not an exclusive, way of thinking in terms of domains. The list of sectors suggested is considerably extended from those conventionally recognized. This extension is indicative of the transition to post-industrialism (c.f. chapter 10).

Table 1 Problem-Oriented Research Domains

Domain	*Key Programme Areas and Aspects*
Medicine	Biological sciences related to medicine; biomedical engineering; clinical and epidemiological studies, including psychological and social aspects; design and appraisals of health care systems and services.
Agriculture	Agricultural sciences and technology; the diffusion of improved practices; the rural economy; psycho-social studies of the changing rural society.
Natural resources	Conservation; recreation; economic aspects; earth sciences; oceanography; social and political regulation; pollution control.
Space	Relevant sciences and technology; uses of space for society; analysis of emergent political issues.
Technology and Industry	Several sub-domains would be required, type of technology giving one possible basis—constructional, mechanical and automotive, electrical, chemical, electronic, nuclear; but technological considered in relation to economic, market, organizational, and human aspects—these are socio-technical systems; relation of public and private sectors, market and social costs.

Table 1—*cont.*

Domain	*Key Programme Areas and Aspects*
Human Resources	The development and deployment of the individual educationally, vocationally, etc.; the educational and employment and career systems and their linkages at all phases of the life cycle; relation to leisure.
Family and Household	Relating the biological, psychological and sociological aspects with those of the economic and material environment.
Community and Regional	Similar aspects at the community level of analysis, whether urban or rural, local or regional; the concept of the 'built environment'—the relation of physical to social planning.
Law and Society	Linking legal, sociological and psychological studies of social regulation in all fields; legislation, courts, police, offenders, prisons, and rehabilitation, etc.; civil, industrial, matrimonial law, etc.
Developing Countries	Cultural, linguistic, racial, economic problems, etc.; population, food, technological transfer; the multi-dimensional nature of the development process; relations with advanced countries.
Advanced Countries	Including the whole range of 'international studies': political, legal, economic, cultural, technological, organizational, etc. Cross country comparisons; the multi-national corporation, trans-national bodies of all types.

Such a listing may serve to disclose the multiplicity and pervasiveness of problem-oriented research domains in the scientific enterprise of a modern society:

The most clearly identifiable and most commonly recognized domains are in the first group centred on a concern with *the resources of the biological and physical environment*—their discovery and scientific explanation, their cultivation, utilization and conservation. Medicine and agriculture are the time-honoured members. Natural resources have belatedly come to be regarded as a comprehensive domain; and now there is space. In most developed and some developing countries, whatever their political systems, there have been created bodies, governmental, private or mixed, which overview these domains.

A review of the British Research Councils and their changing terms of reference would make an invaluable study in this context, especially if compared with analagous bodies elsewhere—not only in the West but in Eastern Europe, where some of the Departments as distinct from the Institutes of the Academies of Sciences perform not dissimilar functions.

The next group centres on technology in relation to industry. But in this vast territory concerned with *the productive capability of the economy* (which gave rise to the classical meaning of applied research) there would appear to be no commonly recognized set of domains. In some European countries technology is perceived as belonging with basic science rather than industry and grouped with it in a Ministry of Science, whereas in Britain the Ministry of Technology and Board of Trade have recently been grouped together, still further emphasizing the separation of applied from basic science which is associated with education in a Ministry of Education and Science. Each of these appreciations has its own validity but the choice made is likely to have far-reaching effects—of a kind difficult to ascertain. Research which is concerned with discovering new or improving existing products or processes may be classified as applied. Such work is related to the world of specific commodities, of market costs and opportunities. These, however, are making unmanageable the world of social costs, many of the effects of economic growth being dysfunctional for the quality of life. It is recognition of this which seems likely to increase the degree to which technological research becomes fashioned in terms of domains. Such domains would bring together the human and organizational, as well as the economic and technical, aspects of productive enterprise and consider its interdependence with other sectors of society. They would thus provide a context for examining the consequences of positive feed-back in what Vickers (1970) has described as a self-exciting system. He has now (op. cit.) introduced the terms, 'user-supported' and 'public-supported' to replace the more conventional designations of the public and private sectors. Ownership of the means of production is not the modal criterion in terms of which relations in this domain, critical for science policy, require appreciation, but rather that of the relation of technology to the quality of life.

The third group, centred on the social sciences, is directly concerned with the *quality of life within a society*. This has become a prevailing topic in the contemporary world because the second industrial revolution, based on the computer and automation, has simultaneously provided opportunities for its enhancement and threats of its reduction which have not existed before. Domain appreciation is likely to be guided by the recognition of two principles discussed in chapter 10:

(i) that the quality of life is affected by the quality of the social reality at all system levels (not merely the individual) and in all dimensions of value (not merely the economic);

(ii) that welfare and development have become inter-dependent in the transition to post-industrialism—welfare in its widest connotation of well-being and functional intactness (stability) and development in its widest connotation of growth (change) that is progressive and order-producing rather than regressive and disorder-producing.

The last group brings together the three previous groups when societies as wholes, or in any of their major aspects, are compared with each other. This they have increasingly to be in a world which has reached a new level of inter-relatedness. Because such comparisons emphasize the culture of particular societies in their historical context this group brings together the social sciences and the humanities. It also involves the *identification and appraisal of trans-national processes and institutions*—a new dimension in 'the order of social magnitude'. This can only be done 'geo-centrically' if multi-national teams are involved in the scientific enterprise, as much as in the multi-national corporation, as Perlmutter (1965) has emphasized.

A first annotation suggests that the relationship between science and society expressed in a problem-oriented research domain is a sensitive indicator of prevailing concerns and values. A scrutiny in terms of resource allocation would confirm this. But the list is incomplete. The area which has consumed more scientific resources than any other among the nations on the winning side in World War II has been left out: *defence*. This is the area which gave rise to the concept of mission. One may question how far defence research has been domain- rather than mission-oriented. Less, perhaps, than one might be inclined to assume, focussed as it has been on weapon systems rather than probable forms of war, the recognition of their true nature, their aetiology and their prevention (Erickson, 1970). Study of these topics has been too much left to small groups on small budgets concerned with peace research and conflict resolution. One may ask also how far defence has dominated the whole field in concept generation as well as in resource consumption, delaying recognition of the distinction between domain and mission.

Despite the ambiguities revealed, there is accumulating evidence that field-determined, generic, problem-oriented research expresses the critical relation between science and society in the transition to post-industrialism. It appears to be so in Eastern European countries as well as in the West; the Academies of Sciences, for example, have formed 'problem-councils' in conjunction with Planning Commissions and State Committees on Science and Technology. This is so also in developing countries making the transition from pre-industrialism to industrialism under the same turbulent conditions in which the advanced countries are concerned with the transition to post-industrialism. A recent UNESCO survey has documented this theme as a world trend (Trist, 1970).

What seems to be required for mapping purposes is a matrix which would relate individuals and supra-individual entities to the main dimensions or sections of a society. An 'entities' list might read as follows:

individuals
families
formal organizations
communities
regions
countries
supra-national institutions.

A dimensions list would comprise the social systems belonging to such sectors as:

education
government
health
industry
law
welfare, etc.

Such a scheme would permit the examination of how far research on health questions was at the individual or the community level, or how far questions of industry and employment were being looked at country-wide or regionally.

More than one matrix would be required:

(a) A useful one could be constructed in terms of prevailing meta-problems which would cut across the categories of sectorial analysis.

(b) A complementary analysis would be in terms of emergent social processes and values discerned by such methods as suggested by Emery in chapter 3 and made necessary by the pace of environmental change.

The three approaches suggested—sectorial systems, prevailing meta-problems and emergent social processes—represent three ways in which problem-oriented domain-based research might be organized. This does not yet provide a theory of such domains though it makes a start in delineating the field.

A Comparison of Domain Selection in Two Countries

The extent to which current science policy is already being implicitly influenced by the domain concept may be illustrated by a brief comparison of science planning in an Eastern and Western European country—Czechoslovakia and France respectively.

In Czechoslovakia a state plan for scientific research 1961-5 was worked out by the Academy of Sciences, the State Commission for the Development and Coordination of Science and Technology (SCDCST), the State Planning Commission and the Ministry of Finance (UNESCO, 1965). The plan was presented in terms of sixteen 'complex projects', some of which represent directions of basic research, while others are problem-oriented. Each complex project was divided into a number of 'fundamental projects', of which there were ninety-five—in turn divided into 'main problems', of which there were 370. These constitute the basic planning units. Work on a main problem is shared by several research establishments, each of which tackles a 'partial problem'. For each complex project a 'collegium' of 8-10 members is appointed, which is the appropriate collegium of the Academy if the project is in an area of basic research. Otherwise, a new collegium is created. If the project is more technological than scientific, the collegium will be responsible to the SCDCST, though it may still be headed by an academician. Its members include chairmen of its fundamental projects and others drawn from appropriate academic, administrative and political fields. The collegium sees that the political, economic and cultural objectives of the plan are attained and appraises overall performance. For each fundamental project there is a more expert body to evaluate and improve the quality of solution of the main problems comprising the fundamental project. It endeavours to keep these in step. For each main problem a co-ordinator links the establishments concerned with its execution and the council of the fundamental project.

Of the sixteen projects in the 1961-5 plan three are centred on the social sciences:

(a) **The role of school and education** in the 'transition from socialism to communism' (equivalent of the transition from industrialism to post-industrialism):—fundamental research in the psychology of learning and the sociology of education; a philosophical and sociological analysis of educational theory; problem-oriented research into methods of instruction and curricula design; and on the relationship of education, training and occupations, including studies of the biological and physiological aspects of work.

(b) **Conditions affecting the performance of the economy** during the transition:—a variety of projects in theoretical, institutional and applied economics; work in industrial sociology and organization theory, a component concerned with the social function of science.

(c) **Social and cultural change** during the transition:—a mix of sociological, political and philosophical studies with contributions from history and ethnology.

In addition, several of the other complex projects involve the social sciences in collaboration with the biological or physical sciences:

(a) **The healthy development of new generations**:—ecological and social factors in physical and mental development, with attention to the psychiatric aspect.

(b) **Nature conservation and healthy natural environments**—a co-operative effort between urbanists, technologists, biologists, sociologists and economists.

(c) **Improved material and cultural standards** through the greater social effectiveness of capital investments:—studies in building economics, town planning, etc.; and an analysis of cost-benefit criteria.

(d) **The automation of complex systems:**—fundamental work on information theory and its application to the social sciences, as well as the social and psychological consequences of automation.

The generic field-determined nature of these projects compels inter-disciplinary collaboration first among the social sciences and then between them and the biological and physical sciences.

In France a more explicit social science policy is in operation under the Ve Plan than elsewhere in Western Europe (DGRST, 1965). The objectives, priorities and instrumentalities of a strategic programme for accelerating the development and application of the social sciences have been systematically worked out and implementation begun. The overriding objective is to build up a comprehensive understanding of the totality of factors affecting economic and social development—psychological, sociological and biological as well as economic and technological. The posture is multidisciplinary and problem-oriented. The aim is to provide a social science capability which will influence national policy. This cannot be done, as the planners see it, by narrowly conceived short-term projects but by thematically conceived and broadly coordinated long-range programmes. The priorities can only be established in the action frame of reference.

Four principal 'orientational' themes have been selected and several programme areas identified in each:

(a) Under **processes of economic and social development** one programme will examine the conditions and consequences of technical innovation and another will consider the relation of standards of living to ways of life. A third will be concerned with the administrative capabilities which facilitate development, and a fourth with urban and regional problems.

(b) The **development of human resources** includes one broad programme area in manpower studies and another on psycho-social aspects.

(c) **Education** is given major emphasis with programmes in learning and motivation; new teaching methods; curriculum content in relation to the obsolescence of knowledge; and the training of future teachers.

(d) The **mutual understanding of societies** is concerned with problems of communication by all methods and media and at all levels of human interaction; also with the conditions of socio-cultural 'equilibrium' and 'disequilibrium'.

To balance this total programme is another concerned with 'free' fundamental research. This is centred on the humanities rather than the social sciences though logic and mathematics are singled out. Additional are the 'Action Concertées'. The first of these is an intensive study of R & D establishments. Others are on programmed learning and the use of space in cities. Finally, there is a scheme for examining neglected aspects of communication and adaptation in groups undergoing changes in life-style through technical and economic development.

In recommending how all this is to be put into effect, the DGRST laid stress on bringing into being special centres in user-organizations, particularly government departments, with the double function of coordinating the projects undertaken and relating the research workers to the executives. The key centre would be set up in the office of the Prime Minister, at the elbow of the Commissariat General du Plan, under the title of the Centre de Coordination d'Orientation des Recherches sur le Development Economique et Social (CCORDES).

The relation of these themes to available resources of money and trained manpower raise awkward questions regarding implementation capability. This is true in both countries as regards funds, but the manpower constraints are even more severe in Czechoslovakia than in France, more especially in the social sciences other than economics. Difficult issues concerning the relevance and effectiveness of decision-making structures in both domain-selecting and research-operating organizations remain unsolved in these as indeed other countries. We are only making a first begining in the field of science policy. These and other questions are beyond the scope of this chapter but are touched on by Trist (1970).

Collaborative Social and Technical Innovation[1]

Social Amplification
The Function of the Professional Relationship
From Project to Programme

Social Amplification

Even if capacity for domain selection at the overall level were well developed this type of science planning could become 'over-determined' if 'spontaneous' engagements between science and society did not also arise from 'below' and then become socially 'amplified'.

The counterpart at the more concrete level of the fusion of social and scientific interests in a problem-oriented research domain is the action-research programme or project carried out by a research group in collaboration with a client system. The client system may be an industrial firm, a public agency in any field or a wider authority embracing a large number of these. The research group may contain any mix of disciplines—whatever is appropriate to the task in hand.

The strategic significance of this type of work derives from the extent to which new institutions have to be built and old ones renewed during a time of social transition as great as the present. Governments now intervene in the operations of society to an extent previously unknown. New schemes need piloting up; their operations monitoring and evaluating. There is a growing demand in the field of social action for an equivalent to industrial development work and also of product innovation

[1] This chapter is developed from a paper on 'Engaging with Large-Scale Systems', an account of the project experience and action-research methodology of the Tavistock Institute, presented to the McGregor Memorial Conference on Organization Development, M.I.T., 1967.

in the sense of the diffusion of a proven pilot throughout the wider system intended. Much of such work is technological as well as social, i.e. socio-technical in the sense of the concept originally introduced in the Tavistock studies of the British mining industry. It involves the joint optimization of independent but correlative systems (Appendix).

The theory of the collaborative relationship and of the practical engagement of social science as a strategy for advancing the base of fundamental knowledge is more advanced than that of problem-oriented research domains. To proceed in the collaborative manner releases processes of social and organizational learning which permit innovations to be accepted and adaptive changes to take place which would not otherwise be possible. The illustrations will be from projects with which the Tavistock group of workers have, with others, been concerned.

The action research studies which provided the first models of the facilitation of planned organizational change as a process involving collaborative relationships between client systems and social science professionals were undertaken during World War II, independently and against the background of distinct traditions, in the United States and Britain when conditions of crisis compelled rapid change. In the United States the main stream derived from the 'real-life' experiments on a number of aspects of social change carried out by Kurt Lewin and his associates (1951). These led to a general formulation of how social change might be facilitated as a three step process (unfreezing the present level, moving to a new level and freezing group life at the new level) which has profoundly affected most subsequent work.

In Britain the counterpart was the development in the war-time Army by a group, most of whom had been at the pre-war Tavistock Clinic, of a form of operational field psychiatry—a sort of psycho-social equivalent of operational research. As the tasks undertaken became more complex psychologists, sociologists and anthropologists were added to the team. Inter-disciplinary collaboration was achieved in an action frame of reference. A common set of understandings developed, based on a shared core value. This represented a commitment to the social engagement of social science, both as a strategy for advancing the base of fundamental knowledge and as a way of enabling the social sciences to contribute to 'the important practical affairs of men'. The value position was the same as Lewin's, though the conceptual background was different—that of a psychoanalytically oriented, interdisciplinary, social psychiatry rather than of a social psychology based on field theory. This core value expresses the underlying assumption on which the work of the Tavistock Institute has continued to be based since it was founded in 1946 to develop further the type of work begun during the period of war. As the British group became better acquainted with Lewin's work its influence on them was far-reaching. The interpenetration of the two traditions was formalized in 1947 in the joint sponsorship by the Research Center for Group Dynamics and the Tavistock Institute of an international journal, *Human Relations,* dedicated to the integration of the social sciences and the continued exploration of the relations of theory and practice as a means of securing their advance.

The method developed had depended in the first place on a free search of the military environment to discover points of relevant engagement. The 'right' had then to be 'earned' to have a critical problem which could not be met by customary military methods referred to the technical team for investigation and appreciation. This appreciation would next be discussed with appropriate regimental personnel and a likely countervailing strategy jointly worked out. The feasibility and acceptability of the plan evolved as well as its technical efficacy would be tested in a pilot scheme under protected conditions and technical control. As the pilot proved itself the scheme would become operational, control being handed back to regimental personnel, the technical team 'retreating' to advisory roles or removing their presence entirely except for purposes of monitoring and follow-up. What was learned was how to take the collaborative role in innovating special purpose organizations with built-in social science capability in a large multi-organization of which the social science professionals were themselves temporary members—the army—under conditions of crisis.

A brief but comprehensive account of the many different types of activity undertaken was published at the time (Rees, 1945). One of the most instructive, from the point of view of engagement with large-scale social systems, was that concerned with the civil resettlement of repatriated prisoners of war. For this purpose twenty transitional communities, designed on data contributed by the repatriates themselves from their own experience, were brought into existence, run very largely by specially trained regimental personnel. *There were never more than two or three psychiatrists in the whole organization.* There were equally small numbers of psychologists and sociologists. Looking back on this experience, what impresses the writer most is not so much the development and maintenance of the transitional communities themselves, which followed on from the two Northfield Experiments (Bion and Rickman, 1943; Main, 1946; Bridger, 1946), as the enormous scale of the effort that had to be made towards the environment, both military and civilian. Without this the necessary sanctioning decisions at the highest level of government would never have been obtained; the repatriates themselves would never have been convinced that the scheme was worthwhile and volunteered to attend Civil Resettlement Units in large numbers; and the participation of some four thousand civilian organizations, especially industrial firms—a quite crucial factor—would never have been obtained (Wilson, Trist, and Curle, 1952).

A recent example is a project on communications in the British building industry begun in 1963 as a joint undertaking of the Human Resources Centre of the Tavistock and its newly founded Institute for Operational Research (Higgin and Jessop, 1965; Crichton (ed.), 1966). The initial relationship was with the National Joint Consultative Council of the Royal Institute of British Architects, the Royal Institute of Chartered Surveyors and the National Federation of Building Trades Employers, to whom the problem has been brought through a resolution having been passed in one of its own branches. The Council then approached the Institute having decided that the problem was one which required social research whose relevance to its needs it had already begun to explore. A small steering committee was set up to work closely with the research team during the pilot phase of the project. In the course of their joint meetings the idea was conceived of reporting back the initial findings to a residential conference, designed on small group principles, convened by the Council and attended by some eighty key influentials from all sections of the industry, the professions connected with it, and the trade unions and government departments concerned. This event took place some six months after the inception of the pilot phase and was the first occasion on which a comprehensive gathering of those having to do with the industry had taken place—its basis being far wider than the Council itself.

The conference, held in a Cambridge College, committed itself unanimously not only to further support of the particular project in question but to take the first step towards setting up an industry-wide research institution to promote a variety of projects in an industry notable for its lack of research interest and now facing rapid technological change and growing crisis in its relations with society. Trustees were appointed and a considerable sum of money raised. A representative research committee was brought into existence which in turn delegated certain of its members to work closely with the research team. In such a steering group key problems can be identified and likely emotional reactions experienced under protected conditions.

Some eighteen months later the pattern of forces in the wider environment had changed in ways that could not have been anticipated, partly by actions of the outgoing government, partly by uncertainties created by the incoming government and partly by changes in the general economic situation. The committee lost any sense of a relationship with a potential comprehensive support base and the trustees felt unable to raise additional funds. The resulting anxieties and conflicts were 'worked through' in the small group composed of the leaders of the research team and the steering committee so that damage was averted to the long-range institution building aims—the strengthening of the industry's general research capability and the development through this of greater powers to collaborate among its conflicting and dissociated interest groups. A 'realization' report was produced of what had been done so far, a shorter version of which the Committee published with a restatement of the aims of the Cambridge Conference. The circulation of both this and the report on the pilot phase have been unusually large, overseas as well as in the U.K. While one or two small project activities have continued on Government research funds, some wider impact has also been made. Moreover, the options have been left open for a joint search to be undertaken to find a new basis for continuing the innovative concept which represented the commitment of the Cambridge Conference. Meanwhile, one of the key ideas—the Analysis of Inter-connected Decision Areas (A.I.D.A.) has been taken into policy research on local government (Friend and Jessop, 1969) and into fundamental research in engineering design (I.O.R., 1967).

A further example is the Industrial Democracy project in Norway (Emery and Thorsrud, 1969) which has now been proceeding for some seven years as a collaborative enterprise between the Norwegian Confederations of Employers and Trade Unions and the Trondheim Institute of Industrial and Social Research and the Human Resources Centre of the Tavistock Institute. At a later stage the Norwegian Government joined the consortium of sponsors while the Trondheim Institute had to set up a new centre in Oslo and the socio-technical group at the Graduate School of Business Administration of the University of California, Los Angeles, has also been drawn in. As with the Building Project this project has required a sustained effort in institution building so that every step would be sanctioned not only by those directly concerned but by those potentially concerned.

The problem arose because of a sudden increase in the Norwegian trade unions of a demand for workers' representation on boards of management. What is remarkable is that the two Confederations should have requested the assistance of social scientists in order to gain a better understanding of what would ordinarily have been treated as a political problem. A thorough analysis of the economic, cultural and political features of Norwegian society was necessary as a background. Since the Norwegians needed to relate themselves to the experience of other countries it is doubtful if a solely Norwegian team would have been credible, but it is certain that a foreign team would not have been acceptable except in relation to a Norwegian Institute which had earned the right to be trusted with such an explosive problem.

The first phase of the project involved a field study of the main enterprises in Norway which included workers' representatives on their boards. The findings, having been reported back to the joint steering committee set up by the sponsors, were widely discussed not only throughout the two Confederations but in the press. The redefinition of the problem obtained in the first phase set the stage for the second which has been concerned with securing, through socio-technical experiments, improved conditions for personal participation as 'a different and perhaps more important basis for democratization of the work place than the formal systems of representation'. The third phase, recently begun, is concerned with the diffusion of organizational learning from these experiments.

Since this diffusion would be mediated through consulting engineers into the bulk of Norwegian industry which consists of small firms it was necessary to find a means of entering the consulting engineering system—through an engineer who had found his own way to a socio-technical approach—Dr. Louis Davis of UCLA. Moreover, the diffusion process has been assisted both in Norway and other countries, where similar socio-technical innovations are now proceeding (Sweden, Eire, and the U.K.) by arranging an interchange of visits by managers and shop-stewards of the firms concerned and also by officials of the relevant trade unions.

The idea of applying the methods of science to social development and innovation is now spreading in a number of different contexts, though it still encounters profound resistance. Russell Ackoff (1968) has summed up the position as follows:

In the democratic societies with which I am familiar, there is almost an innate abhorrence of social and economic experimentation. We think it demeans the subjects and threatens with the possibility of excessive public control of private lives. Yet, curiously enough, no other type of society manipulates and varies the form and content of its control over its members as much as a democratic one. Democratic nations constantly change taxes, tariffs, interest rates, zoning rules, laws, regulations, transportation and communication systems, metrics and even the clock. The major aspects of experimentation—manipulation and control—are already widely practiced in such societies. They even attempt to measure the effects of changes in public policy on national performance. But here is the rub: they usually do not let the design of the evaluative procedure affect the way the public is controlled or manipulated. The evaluators are called in *after the fact,* when it is too late to do an adequate job of evaluation and when the possibilities of gaining understanding are almost completely destroyed. For example, in industry we have found that no amount of retrospective analysis of advertising and sales can yield as much understanding of their relationship as can even a very simple experiment in a few market-places.

In the United States we have just missed a marvellous opportunity to

> perform useful experimentations in connection with our so-called 'Poverty Programme'. Instead of designing these programmes as experiments to inform us how to reduce or remove poverty, we assumed we knew the answers. Only when the failure of such programmes was obvious was any effort made to determine what their effect had been. By then it was too late. Instead of changing our methodology we have only changed our programmes. There is little consolation in knowing that we won't make the same errors twice.
>
> Science and other subsystems of our nation-system must become the subject of experimental study . . . . If experimental designs are used as a basis for the allocation of national resources to science and technology, feedback can lead to adaptation, and gradually improving policy making can be expected while basic understanding of science, if not the nation, is being accumulated.

The increasingly unregulable world which science has been bringing into existence can be brought back into a more regulated state by applying the methods of science to the change processes occurring within it. This entails engaging in social or operational experiments of many different kinds but always sanctioned by those concerned. In this way errors and unintended effects are more likely to be picked up before it is too late. So far as all participate all will learn and so far the values of science will be diffused into the society which will itself embody the social aspects of its science policies.

The Function of the Professional Relationship

This will not happen unless the social aspects of science become embodied in the scientist himself. This comes about so far as he takes professional as well as scientific responsibility for what he does. The professional relationship is differently structured in the social as compared with the natural sciences. It is of special importance to them as a strategy for advancing the base of fundamental knowledge. As this has not been adequately recognized it will be stated in full.

In a paper of the Council of the Institute, 'Social Research and a National Policy for Science,' (1964) prepared as a contribution to the debate which led to the establishment of a Social Science Research Council in Britain, we felt that we should state as explicitly as possible our reasons for assigning central importance to the taking of a professional role in research on organizational development. We had been perceived as having followed too narrowly and applying too generally a clinical model derived from the doctor-patient relationship. In fact, the doctor-patient relationship represents a special case of a very general difference in the relationship of pure to applied science as between the social and the natural sciences. The general argument may be presented as follows.

In the natural sciences, the fundamental data are reached by abstracting the phenomena to be studied from their natural contexts and submitting them to basic research through experimental manipulation in a laboratory. It is only some time later that possible applications may be thought of, and it is only then that a second process of applied research is set under way. The social scientist can use these methods only to a limited extent. On the whole he has to reach his fundamental data (people, institutions, etc.) in their natural state, and his problem is how to reach them in that state. His means of gaining access is through a professional relationship which gives him privileged conditions. The professional relationship is a first analogue of the laboratory for social sciences. Unless he wins conditions privileged in this way, the social scientist cannot find out anything that the layman cannot find out equally well, and he can earn these privileges only by proving his competence in supplying some kind of service. In a sense, therefore, the social scientist begins in practice, however imperfect scientifically, and works back to theory and the more systematic research which may test this, and then back again to improved practice. Though this way of working is well understood in the case of medicine, it is not so well understood that the same type of model applies to a very wide range of social science activities. The model may be called the 'professional model'.

Yet it is often contended that the social sciences should first proceed in isolation, and then one day, when they have advanced far enough in their basic work, they will become capable of developing practical applications. It is further contended that this first stage will be carried through by a few brilliant minds (as it was in the natural sciences) and that, until this stage is over and practical utilization of firmly established knowledge becomes possible, little is to be expected. It follows that for the time being light support only is warranted. Nothing could be more misleading than this viewpoint, which treats the relation between the pure and the applied in the social sciences as comparable with that in the natural sciences, ignores the distinctiveness of the professional model, and fails to see its special relevance to the social sciences.

Another advantage of following the professional as compared with the pure-applied model is that it allows the problems studied to be determined to a greater extent by the needs of the individuals, groups, or communities concerned than by the social scientist himself. This is not, of course, to say that presenting symptoms are accepted at their face value or that the client's prescription is obediently followed. The professional role implies interpretation and redefinition. It does, however, imply a considerable degree of 'field determination'.

Such an approach is proper when a science (and this is the case in many areas of the social sciences) has not yet advanced to the point where there is a large body of fully attested empirical knowledge related to generally accepted theories. For if at this stage the problem is determined too exclusively by the scientist himself, the hypotheses to be tested will tend to be doctrines rather than true theories, or, as a reaction against this, investigation will become artificially restricted to what can be measured exactly. One may expect both too much formal conceptualization of a shallow kind and too much secondary manipulation of meagre primary data.

There is a great difference between simply acting as a consultant and acting as a researcher in a role where professional as well as scientific responsibility is accepted. In the first case there is no commitment to the advancement of scientific knowledge, either on the part of the consultant or on the part of those for whom the inquiry is being made. In the second case this commitment is fundamental and must be explicitly accepted by both sides. It is this that makes the relationship truly collaborative. Though far from all social science research needs to follow such an approach it is unlikely that the study of change processes, and of dynamic problems more generally, can be extended without it.

First attempts to provide a rationale for a professional type of research strategy in the social sciences relied rather heavily on clinical precedents. Recently, fomulations have been proposed, such as that by Churchman and Emery (1966), which are in general terms. After considering the advantages and limitations of traditional 'pure' and 'applied' models, they continue:

A third approach to the study of organizations is to regard the researcher as a member *pro tem* of a third organization sufficiently greater than the organization under study to encompass the conflicting interests and yet sufficiently close to it to permit its values to be related to the concrete issues of conflict. Ideally this third organization would be sufficiently broad to encompass also the interests and values of the research community. However, this implies such a general level of human organization—almost certainly supra-national—that it is difficult to understand how one would work back to agreed-upon objectives in concrete conflict situations. It might be that there are not enough in-between levels at which sufficient community of interests and values exists to justify search for agreed research objectives and criteria. This seems unlikely. Societies as admittedly full of conflict as ours, could hardly hang together unless there were very pervasive strands of common interest. Our own experience is that community of interest can usually be found at the next higher level of social organization. The practical difficulty is more likely to be that the researcher pursuing this third approach will have to engage in institution-building so that agreement about his research concerns can be actively pursued and powerfully sanctioned. It should be noted that when in this approach the researcher obtains his value standards from the next higher level he does not have a privileged objective standing as a member of this level. He can claim neither special knowledge of the value structure of this level nor special power to sanction things on its behalf.

From Project to Programme

This approach has particular relevance to situations in which the research team has to engage with large social systems involving sets of organizations rather than a single organization (c.f. Evan, 1966). These inter-relations comprise systems of *organizational ecology*. During the decade of the 'fifties' this broader contextual element was absent in the action researches in which the writer and his colleagues had opportunities to participate, as was the deeper psychological resonance earlier evident (Bion, 1943, 1947-51). Collaborative studies, in response to the changed character of client demand, focussed rather sharply on the single organization and those parts of society and people which this renders visible. An animated search took place for a 'new learning' about 'organizational-life', which produced a fundamental revision of the concept of bureaucracy. This was the needed outcome for increasingly complex organizations facing increasingly complex environments. While the advances made will be carried forward, the path-finding projects of the next few years are likely to be characterized both by a greater social extensiveness and a greater psychological intensiveness than the major collaborative studies of the 'fifties'. It is scarcely accidental that an ecological re-appraisal of organization theory by Vickers (1959, 1965, 1968) and a 'socio-experiential' re-appraisal of personality theory by Laing (1960, 1961, 1964, 1967) should have been proceeding contemporaneously.

In endeavouring to understand and respond to the changing social environment of the decade of the 'sixties' workers at the Tavistock have found themselves undertaking projects with the following characteristics: the domain of concern is with multi-organizational clusters rather than with single organizations; this has had the effect of directing attention both to the wider society and to the individual as a member of the social aggregate. Such projects have a future orientation. They increase the time-scale over which social science resources must be committed and extend the nature of the inter-disciplinary mix. The international ramifications also become greater, both as regards the countries offering suitable fields of study and the composition of research teams.

If the need is for the social scientist to become directively correlated with the meta-problems that permeate Type 4 environments, collaborative research must be developed at the programmatic rather than the project level. Projects relate the research worker to specific rather than generic problems. These problems are experienced as the client's rather than his, whereas a meta-problem is experienced as his also; it belongs to both as members of the wider society which it permeates. It is for this reason that the collaborative researches of the next few years are likely to be characterized by a greater psychological intensiveness as well as by a greater social extensiveness. If the latter makes the single organization no longer appropriate as the unit of analysis, the former imparts an existential quality to the work undertaken.

To the writer it feels as though he 'has been here before'. This feeling of *déjà vu* made him recall the type of work in which he participated during World War II. At the present time he experiences those with whom he works as travellers on a common journey rather than as clients who have requested his professional help (c.f. Laing, 1967).

The identification of themes for programmes cannot in the case of collaborative research be made in the abstract. These themes can be reached only by an analysis and realization of the nature and meaning of field experiences which carry social science engagement. During the summer of 1966 the members of the Human Resources Centre of the Tavistock Institute held an off-site conference to review their recent work in order to improve their understanding of its underlying thematic content. A summary of our findings is offered below. This may exemplify the kind of assessment which many research groups are likely to feel a need to make in order to improve their capability for programmatic engagement.

The first step was to identify programme areas. These represented domains of action research in which there was a convergence between increasing preoccupation among client organizations and persisting scientific interest among staff members. Four were identified which comprised a frame of reference within which research initiatives could be taken and approaches by client organizations appreciated:

(1) Creating non-alienated work relationships.
(2) Life careers and the changing character of work.
(3) Institution-building in complex and uncertain environments.
(4) Diffusion of social science knowledge and capabilities in user organizations.

Projects tended to have their origin in problems of alienation; from this arose concern with the effect on individual lives and careers; and thence a need to build new social institutions and to diffuse the knowledge gained.

Creating Non-Alienated Work Relationships

Earlier socio-technical studies had shown that human needs, satisfactions and interests could be met in the work situation without sacrificing economic goals. Work alienation was not a necessary consequence of attempts to increase economic efficiency. Current interest lay in designing jobs and organizational forms so that alienation could be replaced by involvement.

Life Careers and the Changing Character of Work

A noticeable shift had taken place in many organizations from a pre-occupation with filling 'job slots' towards a concern with the long-term planning of careers. A new objective was the development and release of human resources. There was a need to consider the life-strategies of the individual considered as the career-carrier in relation to the crises of the life cycle and the changing position of work-roles in the life space.

Institution-Building in Complex and Uncertain Environments

Re-analysis of experience with a number of large organizations had suggested that many of their common problems arise from the need to cope with environments of increasing complexity and uncertainty. The study of what is involved in building institutions better able to meet this 'turbulence' had been hampered by assumptions of the closed system theory of organization. Work was now beginning to centre on problems of management philosophy and company objectives, and the identification of the values the organizations needed to pursue if they were to manage themselves in increasingly complex and uncertain environments, and to build these findings into organizational practice.

Diffusion of Social Science Knowledge and Capabilities in User Organizations

This developing area was concerned with discovering general principles which govern the acceptance and diffusion of the types of change that depend on the use of the insights and skills of social scientists by those who are not social scientists. Experience of the complexities and unpredictable positive and negative outcomes encountered in attempts to develop competences and understanding gained from field studies in Norway, Eire and the U.K. indicated the urgent need to develop greater knowledge of the diffusion process. Recent work concerned with change, leadership and diffusion had centred on the development of forms of task-oriented group relations training appropriate to the development of a greater understanding of organizational processes, values and objectives.

The conditions postulated as requisite for realizing this total programme may by summarized as follows:

(1) Long-range collaborative relationships covering work in all four areas must be established with a set of enterprises in science-based industries of sufficient

stature in the British environment to command respect in the country as a whole, and therefore to be capable of setting an example.

(2) The trade-unions concerned must participate as well as the managements.

(3) The enterprises must agree to work with each other as well as with the Institute so that comparisons could be made and seminars, workshops and labs held across the set.

(4) Independent research funds must be secured as well as funds from client sources. This step had been requested by several client organizations as well as by the Institute, so that sanction could be secured at a public level, giving initial protection for the risk of innovation yet enhancing the chances of later diffusion.

(5) The set of studies under British conditions should be complemented by studies in other countries, not only to provide international comparisons, but because the environments of some countries provide opportunities not available in others, and because interchange of managers and trade unionists between collaborating enterprises in different countries had already proved a potent method of change induction through mutual learning.

(6) The Centre should commit itself to the programme as a group enterprise, although all members would not necessarily be involved whole time. Some would provide essential links with other programmes.

(7) The Centre should not attempt to carry out the programme solely with its own resources. Therefore arrangements of a selectively interdependent character would have to be increased to secure personnel (senior as well as junior) from other research organizations both in Britain and in other countries.

(8) These arrangements should be two-way. Therefore, some members would always be out-posted to collaborating Institutes and Universities, sometimes taking up permanent appointments, whole or part time, with these bodies, just as permanent staff members were received from them.

(9) An international post-graduate training responsibility

was recognized in providing fieldwork opportunities for doctoral candidates. The participation of such students should be accepted by client organizations wishing to enter a fully collaborative relationship.

(10) The inter-disciplinary mix should be extended. The main developments so far had been the involvement of the Institute for Operational Research, a member organization of the Tavistock Institute, and of engineers from the U.S. and Norway, and of economists and political scientists from the U.S.

This entire set of conditions has been obtained on a long-range basis for the Norwegian Industrial Democracy Project in which the Centre collaborates with the Institutes of Work Research in Oslo. Apart from the period towards the end of World War II, when innovations such as the Civil Resettlement Units were in process, the entire set of conditions has been approximated in our work in Britain only temporarily for such schemes as the programme concerned with Communications in the Building Industry.

In larger and more complex societies change processes involving social as well as technical innovation and embodying new values and practices in key organizations are likely to be more difficult to generate and sustain than in smaller societies whose cultural traditions lend themselves to certain lines of development of general significance for the transition to post-industrialism. Many part-processes may begin in the larger countries, but few are likely, especially within a relatively short time-span, to attain 'critical mass'. In a very large society such as the United States exciting 'starts' can occur in innumerable settings but the scale and heterogeneity being of another order from medium sized European countries, the difficulties are even greater of adaptive innovations taking a firm hold. Small homogeneous societies may under some circumstances remain change resistant to a greater degree than large heterogeneous societies. When, however, a particular process sets in which their culture enables them to carry forward, a new pattern can establish itself rapidly and extensively in view of the connectedness of the networks in and around the 'leading part'. Such societies may therefore develop a special role as 'social laboratories' in the transition to post-industrialism. Other countries may learn much from their experience of inaugurating developments with which they have been unable to make headway themselves but which are relevant to their future.

Chapter 10

The Relation of Welfare and Development: Systems and Ideo-Existential Aspects[1]

The Post-Industrial Society
Some Key Background Concepts
Systems Aspects
Ideo-Existential Aspects

The Post-Industrial Society

The degree of change now taking place in the contemporary world is of an order as great as that occurring when large-scale societies with written languages first arose on the basis of agricultural settlements. This ushered in what Kenneth Boulding (1966) has referred to as the era of 'civilization' which, having lasted some 5,000 years, is beginning to give place to a new type of social order. Noting that the most critical single change during the period of civilization was the transition from pre-industrial to industrial societies, Daniel Bell (1965) has won acceptance for the term 'the post-industrial society'—originally suggested by Riesman (1958)—to indicate the character of the emerging society. In post-industrialism the available technology will no longer absorb the bulk of the energies of most of the people.

Though post-industrialism in a fully developed state is unlikely to be reached anywhere in the world for a considerable number of years, the existence of an irreversible trend in this direction is already powerfully affecting ever widening classes of events and larger masses of people, both consciously and

[1] This chapter and chapter 11 are drawn from a monograph prepared for the Seminar on 'Conceptual Aspects of Welfare and Development' sponsored by the Canadian Centre for Community Studies, Ottawa, 1967. The original was revised as Working Paper No. 1, Socio-Technical Division, Western Management Science Institute, U.C.L.A., 1968.

unconsciously. This trend is proceeding far more rapidly and far more unevenly than had been anticipated within as well as between countries. If the gap between the developing and the developed countries is widening, so in the latter is that between diverse sets of advantaged and disadvantaged groups. Another widening gap is between generations. In the most advanced countries certain parts of the society are already in or approaching an early phase of post-industrialism, while many others remain in various phases of industrialism, and still others are pre-industrial. Under some conditions these different parts are interspersed. Under others they are separate. With the means of communication now available (especially through television) a diffuse consciousness of this total state of affiars is spreading, making reactions more rapid in unstructured publics and organized interest groups, and altering the threshold of what will be tolerated. The problems created are of a type and on a scale which call more than ever for planned intervention. But too many of the attempts so far made have had poor success. Meanwhile new forms and degrees of violence and estrangement are appearing.

While the rate of technological advance makes the transition to post-industrialism inevitable, immense variation is possible as regards the amount of dislocation and conflict incurred and the 'quality of life' which becomes established. This variation is open to human control to an extent which major social transformations in the past were not, for the requisite knowledge has become available. Though the existence of knowledge does not guarantee its use, a new opportunity is created by its presence.

Some Key Background Concepts

The analysis here to be attempted will be based on a way of thinking about systems and their environments developed from the distinction drawn by von Bertalanffy (1950) between 'open' and 'closed' systems. Its relevance to the large scale populations and systems with which policy-makers are concerned has only recently been appreciated. Lawrence K. Frank (1967) suggests that it is to these ideas we must look for a new political theory. There is emerging a 'Service State', not to be confused with the 'Welfare State' with its aura of charity and philanthropy. The distinguishing characteristic of the Service State is its commitment to enhance the 'well-being' of everyone. Welfare is given its true meaning of well-being, which is to be 'enhanced', that is to say, 'developed'. Welfare and development have become inter-dependent. Both are now being affirmed as universal rights. The perception of their relationship and the assertion of these values is in our view the nub of the matter.

This view is reflected in current statements by leading professional bodies, at the national and international levels concerned with problems of welfare: for Canada, in the papers presented to the Conference on 'A Comprehensive Statement on Social Welfare for Canada' (Canadian Welfare Council, May 1967); for the United States, in the H.E.W. pamphlet, 'Social Welfare in a Changing World' (Wickenden, 1965); for the United Nations, in the '1963 Report on the World Situation' (U.N. Department of Economic and Social Affairs, 1963). These writings indicate that a major advance in welfare thinking has taken place.

Our own analysis will begin by considering certain general characteristics which have become salient in the 'causal texture' of the contemporary environment (chapter 4). This is regarded as taking on the character of a *turbulent field.* This turbulence arises from the increased interdependence of the parts and the unpredictable connections which arise between them as a result of the accelerating but uneven change rate. This turbulence grossly increases the area of relevant uncertainty for individuals and organizations alike, and raises far-reaching problems concerning the limits of human adaptation. Forms of adaptation, both personal and organizational, developed to meet simpler types of environment no longer suffice to meet the higher levels of complexity now coming into existence. Emery (chapter 5) has extended this analysis with special reference to the transition to post-industrialism. Having described a set of 'social defences' which represent maladaptive 'strategies' for complexity reduction, he proposes a principle of organizational design which can allow higher levels of complexity to be tolerated. This principle, which he calls the 'redundancy of functions' in constrast to the 'redundancy of parts', is of central importance to our analysis of the concepts of welfare and development. We shall return to it at a later point.

Meanwhile, Michael Chevalier (1966, 1967) has drawn attention to the new type of diffuse social problem which arises under conditions of turbulent (Type 4) environments, which he refers to as a *meta-problem.* Not only have problems developed far wider ramifications but this quality is becoming more widely perceived. This represents a new factor which is 'existential'. 'Society has come more and more to perceive and articulate a new kind of problem. It is not only a matter of putting related problems together; new knowledge and expectations have caused a fusion, an interrelation of problems into a class of meta-problems. And society, once having perceived a meta-problem, begins also to perceive that courses of action to relieve it are inter-related. In fact, some comprehensive attack is now the only strategy acceptable to society'. Poverty, Environmental Pollution, Regional Disparity, and Bi-lingualism and Bi-culturalism are issues widely recognized in Canadian society as constituting meta-problems. Bertram Gross (1967) lists 'systemic' problems for the U.S. In Type 4 environments activities in the domains of welfare and development become directively correlated with meta-problems. This needs to be realized by those responsible for formulating and executing policy. Otherwise there will be no 'engagement' between political and interest group leaders, agency administrators and the numerous organizations and diffuse overlapping publics whose needs their policies are framed to meet and whose condition their actions are intended to improve. Effective solutions to meta-problems depend on collaboration. Coercion cannot be effectively exercised across the number of inter-faces involved. The higher level of complexity calls for a new mode of adaptation.

The new adaptive capability required must have characteristics which will enable it successfully to contend with the new basic condition: that the meta-problems of Type 4 environments are from an organizational point of view, *ecological.* With the single organization, however large, which usually has one general purpose together with a limited number of more specific objectives reconcilable through compromise, we have become increasingly expert. In handling organizational interdependencies, where purposes are many, and priorities and conflicts less easily reconcilable, we remain, by comparison, novices. These interdependencies and their relation to the unstructured publics which constitute the social aggregate of the overall society create problems of *organizational ecology.* We remain novices because we have been used (except in times of crisis such as depressions and wars) to a society in which, by and large, problems of social ecology have taken care of themselves. This they were expected to do in the ethos of the industrial society, above all by the 'free play' of the market, which was supposed to 'produce' the greatest good for the greatest numbers. But with the increasing salience of Type 4 conditions auto-regulative processes, to use a term of Michel Crozier's (1964), are breaking down. We can no longer depend on them. In the opinion of Vickers (1968) societies on the threshold of post-industrialism are in danger of falling into what he calls 'ecological traps'.

If we cannot depend on auto-regulative social processes to take care of the future, we have to intervene ourselves. The *necessity for intervention* follows from the nature of Type 4 environments and the consequent transformation of problems into meta-problems whose character is ecological. The *object of intervention* is to increase the probability of securing the advent of one of the more rather than one of the less desirable of the 'alternative futures' which seem to be open. The *instrument of intervention* is 'adaptive planning'[2]—the working out with all concerned of plans subject to continuous and progressive modification which are what have to be made when what has to be done cannot be decided on the basis of previous experience. The *agency of intervention* is government—but in collaboration with other key institutional groups—for adaptive planning will require the active participation as well as the free consent of the governed. Nevertheless, governments, whether at the national, regional or local levels or in relation to the sectorial activities and organizations which traverse them, are the bodies whose function it is to maintain a society in some kind of ecological balance. *The function of a new politics will be to establish the enabling conditions for this balance to maintain itself.*

Emery (chapter 6) has stated the problem in terms of the need to take an active rather than a passive role and has drawn on the theory of directive correlation formulated by Sommerhoff (1950, 1969) to distinguish between active and passive modes of adaptation. Though made in terms of the need for the social scientist to adopt an active role in widening the

[2] For this concept see Ackoff, R. L., (1969). A Concept of Corporate Planning.

options which decision-makers can take into account, the argument holds in the general case. In the limit, all members of a society must in one way or another be involved in taking the active role, if the passive role is no longer available. *The policies devised and the programmes undertaken when an active mode of adaptation is pursued at one and the same time represent an effort of development and a strategy of welfare.*

Systems Aspects

We shall now remove the concepts of welfare and development from their everyday connotations and attempt to establish their basic characteristics as general properties of open systems. This will entail (a) distinguishing the system states to which each refers; (b) considering additional factors which must be taken into account in relation to social as compared with biological systems; (c) similarly, with reference to larger rather than smaller social units; (d) explicating the relations of the two concepts to each other under different sets of conditions; (e) this last with special reference to the implications of system complexity when associated with rapid and uneven change.

In the relations of the system (whether organism or organization) to its environment welfare and development are complementary states which are positive for the adaptive process. They have thresholds (standards) which must be determined empirically. The attainability of states above the threshold raises a further set of questions which need not detain us here except to note that in the case of man the limits remain unknown. Below the threshold welfare and development turn into their opposites–states which are negative for adaptation and survival.

(a) *Welfare,* or *well-being* (to continue to function well), refers to states of a system under conditions which maintain the steady state. Its opposite, *ill-fare* or *ill-being* (to be dysfunctional), refers to states of a system under conditions which do not permit the steady state to be maintained. This whole set of terms is concerned with the 'statics' of adaptation–with stability (not to be confused with stagnation which is a state of ill-fare) and with the regulation and maintenance of stability.

(b) *Development,* or *progression* (to continue to advance), refers to processes by which a system reaches higher order steady states of a more adaptive nature. Its opposite, *deterioration,* or *retrogression* (to go back), refers to processes by which the system returns to states of a lower order (stable or unstable) which are mal-adaptive. This whole set of terms therefore is concerned with the 'dynamics' of adaptation—with positive change leading to the establishment of widened and preferred orders (not to be confused with negative change which leads to disorder or to greater constraint). Development involves discovery and innovation. It is concerned with the regulation and maintenance of growth.

Table 1 Bio-Social and Socio-Cultural System States

Welfare (Well-being)	*Ill-fare (Ill-being)*	*Development (Progression)*	*Deterioration (Retrogression)*
Intactness	Impairment	Maturation	Arrest
Robustness	Vulnerability	Learning	Retardation
Self-regulation	Breakdown	Extended adaptability	Restricted adaptability
Integration	Dissociation		
Independence	Dependence	Cultural accumulation	Stagnation
Interdependence	Isolation		
Coordination	Scatter	Product accumulation	Waste
Cooperation	Conflict		
		Environmental expansion	Contraction
		Innovation	Obsolence

Both biological and social systems contribute to these outcomes since man belongs to both. Without assuming (Table 1) complete isomorphism of system properties between biological and social systems or between social units of different social magnitudes, we may essay a first list of properties basic to states of human welfare and development and their opposites. As we pass from bio-social to socio-cultural systems, in relation to welfare we may transpose the concept of the intact functioning organism (organization) as among the necessary conditions; but add that of a higher order of intra-population interdependencies as among the sufficient conditions for maintaining the steady state. In relation to development we may transfer the concepts of maturation and learning; but add those of actively transforming the environment (through technological change) and of cumulating information (through cultural transmission). These points are related to the fact that the socio-cultural systems which the bio-social human individual forms with other such individuals have an altogether higher order of openness from those formed by other species. The thresholds (standards) themselves change, as the norms which determine expectations change with society.

In relation to socio-cultural systems, welfare and development share a *common set of dimensions.* These represent *categories of value* such as those proposed many years ago by Spranger (1928)

(a) economic
(b) social
(c) political
(d) scientific
(e) aesthetic
(f) religious.

States of welfare/ill-fare and of development/retrogression exist in some such multidimensional set as universal attributes, however much emphasis may vary between societies and among individuals and groups within a society. Certain thresholds of provision tend to become established as rights through the operation of social norms; certain thresholds of performance as duties. In the 'progression' from pre-industrial, through industrial, towards post-industrial societies, each of these dimensions has tended to establish itself as a domain within which welfare and development rights may be asserted and duties expected. Moreover, the thresholds have tended to be set at higher levels.

Welfare and development share in common *referents* at *all orders of social magnitude:*

(a) the individual;
(b) the family and various forms of kinship system;
(c) formal and informal organizations;
(d) communities, i.e. ecological systems: local, regional, national;
(e) Transnational entities—even the world as an emerging interdependent system.

States of welfare/ill-fare and of development/retrogression exist in social units at all system levels as well as in all socio-cultural dimensions. The levels are qualitatively as well as quantitatively different as regards the types of relationship they involve. Their welfare and development requirements pose problems which are correspondingly different, and which may be in conflict. Nevertheless, they are inter-dependent, the degree varying with the complexity of the environment.

If in pre-industrial societies the *kinship system* tends to be the most salient component of the social structure, in industrial societies it is *formal organizations;* while in post-industrialism it would appear that *ecological systems* are likely to take this role.

The relations of welfare and development take three principal forms: when development is a function of welfare; when welfare is a function of development; when welfare and development are inter-dependent functions. The form of these relations is determined by the types of environment with which a society, or a relatively autonomous part of it, are directively correlated.

Development as a function of welfare. This state expresses the relation which obtains under conditions of the more placid environments where the maintenance of stability is the principal requirement for adaptation. This state is typified by pre-industrial societies, particularly in their earlier and simpler forms:

(i) Welfare is maintained by auto-regulative processes operating through the kinship system, which plays the role of a 'leading', or pivotal, part.
(ii) Development measures are required to maintain established states of welfare when auto-regulative mechanisms can no longer cope in face of internal and external

threats. Development processes under these conditions are not auto-regulative but involve taking the active role. Modes of intervention in these societies are characterized by coercive methods, illustrated in the rise of autocratic regimes and regular armies.

Welfare as a function of development. This state expresses the relation which obtains as the environment becomes more dynamic, when internally generated growth (resulting from technological change) is now the principal requirement for adaptation. This state is typified by industrial societies.

(i) Development is maintained by auto-regulative processes operating through the market system, where enterprises now play the role of the leading part.

(ii) The welfare of increasingly numerous classes of people and segments of the society is no longer auto-regulative. The maintenance of their welfare requires the taking of an active role. Modes of intervention in industrial societies cannot remain solely coercive if disturbances of a revolutionary type are to be offset. Legislative reform based on 'democracy by consent' (Clegg, 1960) makes its appearance.

Welfare and development as inter-dependent functions. This state expresses the relation which obtains as Type 4 conditions become salient in the transition to the post-industrial society. Adaptation now depends on the ecological regulation of the interdependencies in all their dimensions of the innumerable sub-systems which characterize large societies undergoing rapid but uneven change.

(i) The welfare of sub-systems now inherently involves their development; otherwise the accelerating change-rate soon renders them obsolescent—when they fall into states of ill-fare.

(ii) Sub-system interdependence also increases so that states of ill-fare in a relatively few sub-systems (especially if their position is crucial) can produce widespread dysfunction in larger systems. The development of particular sub-systems is dependent on the welfare of other sub-systems to a greater extent than when the degree of societal inter-dependence is less.

(iii) Unevenness in a change-rate widens the range of outcomes, so that social segmentation increases. The number of groups perceiving themselves, or being perceived, in sub-threshold states becomes greater as the expectations which set the thresholds rise and as the sense of 'relative deprivation' grows.

(iv) The effects of these contradictory trends are magnified by an increase in the number, diversity and size of sub-systems, which raises the overall level of complexity.

(v) This, in turn, raises the level of uncertainty. It now becomes less possible for a given sub-system to remain directively correlated with a relatively closed set (e.g. for agriculture to live in the 'rural world'). Each member of the immediate set to which a sub-system belongs tends to be linked with a growing and changing number of other sets, which cannot be completely identified. These sets tend to be related to each other in different ways, and often belong to different 'universes'. It therefore becomes harder to predict if, or for how long, a particular sub-system will continue to develop, or remain in a state of welfare.

The meta-problems created in this situation pass the limit within which auto-regulative processes can adaptively operate with respect to either welfare or development, so that an active role becomes required.

As this need to take the active role becomes more general it changes the quality of the society. Hence the need to distinguish post-industrialism from industrialism. Hence also the relevance of Lawrence Frank's contention that a new political theory is required.

With the passive role no longer constituting an option, the central issue becomes the character of the active role. There are two main and opposite directions in which this role can develop. One, modelled on the principles of the physical sciences, would lead towards a more engineered society. The other, modelled on the principles of the life sciences, would lead towards a more organic society. Emery (chapter 6) has made it possible to state the choice between these two models in system theory terms.

'The choice is really between whether a population seeks to enhance its chances of survival by strengthening and elaborating special social mechanisms of control or by increasing the adaptiveness of its individual members; the latter is a feasible strategy in a turbulent environment and one to which western societies seem culturally biased.

'We have stated that choice is unavoidable. What makes it unavoidable is what we might clumsily call a design principle. In designing an adaptive self-regulating system, one has to have built in redundancy or else settle for a system with a fixed repertoire of responses that are adaptive only to a finite, strictly identified set of environmental conditions. This is an important property of any system, as an arithmetical increase in redundancy tends to produce a log-increase in reliability. The redundancy may be achieved by having redundant parts but then there must be a special control mechanism (specialized parts) that determine which parts are active or redundant for any particular adaptive response. If the control is to be reliable it must also have redundant parts and the question of a further control emerges. In this type of system, reliability is bought at the cost of providing or maintaining the redundant parts, hence the tendency is toward continual reduction of the functions and hence cost of the individual part. The social system of an ant colony relies more upon this principle than does a human system, and a computer more than does an ant colony. The alternative principle is to increase the redundancy of functions of the individual parts. This does not entail a pressure toward higher and higher orders of special control mechanisms, but it does entail effective mechanisms within the part for setting and resetting its functions—for human beings, shared values are the most significant of these self-regulating devices. Installing these values of course increases the cost of the parts. The human body is the classic example of this type of system, and it is becoming more certain that the brain itself operates by means of overlapping assemblies based on similar sharing of parts.

'Whatever wisdom one attributes to biological evolution, the fact is that in the design of social organization, we have a genuine choice between these design principles. When the cost of the parts is low (in our context, the cost of individual life), the principle of redundant parts is attractive. The modern western societies are currently raising their notion of the value of individual life, but a change in reproductive rates and investment rates could reverse this. There is, however, a more general principle that favours the western ideal. The total error in a system can be represented as equal to the square root of the sum of the squares of all the component errors. It follows that a reduction in the error of *all* the components produces a greater increase in reliability than does an equal amount of reduction that is confined to some of them (e.g. to the special control parts). We are certainly not suggesting that this principle has been or is even now a conscious part of western ideologies. Some sense of it does, however, seem to have reinforced our prejudice toward democratic forms of organization.'

Our analysis may now be extended as follows: that at the higher level of complexity which characterizes the transition to post-industrialism a higher quality is required in *all* primary social units. By primary social units is meant the set of concrete social resources which exist in the life-space of the individual, i.e. the people and institutions with which he directly interacts and to which he contributes his own resources: his family, his work-place, the school his children attend, the particular community in which he resides, the services and amenities actually available to him: in sum, all those entities which compose his *primary social world.* The quality of these resources, in his case, determines for the individual his 'quality of life', on which his welfare and development alike depend. The objective of taking the active role is to bring into being ecological systems able to maintain primary social worlds of high quality throughout a society. Entailed are making changes at each system level in the order of social magnitude and in all dimensions of value.

Ideo-Existential Aspects

Concepts such as welfare and development do not originate in the language of science but in the *political vernacular.* They express complex, diffuse and shifting 'images' (c.f. Boulding, 1956) which variously reflect the consciousness and actions of a society. Concepts of welfare and development belong to a class of phenomena which may be termed *ideo-existential.* They have a double reference: to *social ideas and values;* to *social practices and processes.*

The characteristics of ideo-existential phenomena may be summarized as follows:

Ideo

(i) **Cognitive**: ideals, philosophies, beliefs, values, etc.—the entities which compose a social ethos. Such entities represent forces which have the capability to shape the 'appreciation' of meta-problems. Examples are the statements of the Canadian, American and U.N. authorities on welfare and development problems. Such statements help to shape the cognitive structure of a society, providing the categories and axioms in terms of which the world is construed. They are receiving increasing attention from social anthropologists (Barth, 1965).

(ii) **Motivational:** Goals, aspirations, norms, expectations, affects, etc. These transform the 'guiding fictions' of the cognitive structure into activators of behaviour. The motivational situation may be such that no such transformation takes place. It cannot be assumed that because certain doctrines of welfare and development are rational, policies based on them will be implemented.

The relations between guiding fictions and activators are not directly inferrable but must be established empirically. There are, moreover, intervening variables between the cognitive and motivational aspects of ideo-existential phenomena such as national or social character. Dicks (1950) established a relation between Nazi ideology and the political behaviour of Germans through a 'depth analysis' of the German character based on interviews of German prisoners of war. Similarly with his analysis of the Russian character (1952). Hagen (1960) has attempted to relate the Protestant ethic to the behaviour of entrepreneurs in the industrial revolution by an analysis of the social character of English non-conformists. McClelland's work (1961) on achievement motivation is concerned with similar problems over an array of societies. Emery's current work (1968) shows that certain features of the English character which are duty rather than achievement oriented may impede the implementation of policies concerning economic growth:

Existential

(i) **Social practices and processes:** behaviours representing the embodiment in customs, organizations, interest group coalitions, etc. of motivationally effective social ideas. They are evident in court judgements and legislation, and may be followed in administrative rules and decisions and in the direction of social change.

(ii) **Social perceptions:** subjective constructions of these practices and processes by individuals and groups in the societies concerned. These perceptions are shaped in a far-reaching manner by feelings about the relevance of these practices and processes for 'us'; their goodness/badness for 'us'; doubt about their degree of reality/unreality; knowledge/ignorance of them; trust/mistrust of them; confusion and mystification by their frequent inconsistencies (c.f. Laing, 1963); etc.

These perceptions affect operating social practices and processes through the effects they have on the cognitive and motivational aspects. A circular causal process (Lewin, 1951) is continuously at work relating the ideo and existential fields to each other.

Ideo-existential phenomena have a number of general properties which must be taken into account in any analysis of welfare and development problems in a society.

(a) *Various subjective interpretations* of them exist at any one time. This is illustrated in the work of Bott (1957) on subjective models of the class system in contemporary Britain.

(b) *Their salience order is also variously perceived,* as is so frequently exemplified in opinion polls on topical issues, using stratified samples.

(c) There are *various degrees of congruency/incongruency with other such entities* expressing other systems of beliefs and values, especially in large pluralistic societies.

(d) There are *competing claims of other classes of needs and interests,* even if no incompatibility exists in terms of values and beliefs.

(e) They exhibit *varying degrees of diffusion* in groups and institutional sub-systems. They may affect only small minorities or pervade the society as a whole.

(f) They are subject to *differing evaluations,* from the extreme positive to the extreme negative poles, by different interest groups. These may be social classes, 'vested interests', key minorities, or the target groups of government programmes, or the legislating or administering authorities concerned. Many types and degrees of conflict may arise when this condition is put together with those listed above.

Inherently present in ideo-existential phenomena there are two tendencies:

(a) For a gap to exist between the ideo side and the social embodiment side. This may be relatively slight or become a serious '*credibility gap*'.

(b) For a gap also to exist between what is taking place and what is supposed to be taking place. Gross (1967) has called this the '*intelligence gap*'. Ideo-existential phenomena offer immense scope for distortion, delusion, dogma and idealization.

As regards action, there is a similar scope for irresolution of issues, for 'non-starts', for what Selznick (1957) has called *drift;* similarly, for what he has called *adventurism*—crash programmes, opportunistic sallies, etc. There is also scope for what we shall call *retraction,* i.e. withdrawal of commitment, denial of original intent, 'cut-downs', etc.

These gaps between apparent intent and consequential action have reached serious proportions in large pluralistic societies. Very different is the situation in smaller and more homogeneous societies. Modern information technology, especially in the use of mass media such as television, is capable of substantially reducing these gaps. How far it will be used to do so will depend on the choices which a society makes in taking the active role. To secure widespread support for and effective implementation of comprehensive welfare and development programmes will necessitate a substantial reduction of both gaps.

The character of ideo-existential phenomena changes through time. Forms once adaptive become maladaptive as social change proceeds. They become *depassed* (Sartre, 1964). They leave *residues* as Pareto long ago described. They occasion innovations involving the *displacement of concepts* (Schon, 1963). While some ideas, motivations, practices and perceptions disappear entirely, others appear for the first time through this latter process. This consists of refashioning what already exists, though often in separate parts of the social field, and a bringing of it together, a redefining of it, to suit the needs of a new context.

(a) Some key *ideo-existential phenomena persist over long periods of time,* but go through many 'editions', several of which may exist simultaneously, or overlap. Some people are in one edition, some in another. Others do not know which edition they are, or should be, in. This aggravates the problems of the credibility and intelligence gaps.

(b) In a period of profound and rapid social change ideo-existential phenomena tend to *persist in forms opposed to the needs of emerging realities.* While this may be appreciated relatively early by certain elites attuned to detecting areas of likely crisis, it is usually relatively late before this appreciation is wide enough to provide a support base for taking corrective action on new lines.

(c) The *persistence of outmoded forms is unevenly distributed* among the many strata and sectors which make up a large pluralistic society. This is related to the uneven rate of change, and also to the 'make-up' of the various segments.

(d) There is a *widespread 'cultural lag' associated with*

resistance to change. The core values of a society and the basic shape of its institutions are exceedingly difficult to change—alike for conscious and unconscious reasons. Very great anxiety is aroused by the need to change and by the process of changing. The preferred mode of accommodation is to proceed piece-meal and to make the process long drawn out. But the realities do not always permit this. When this is sensed one may expect the appearance of massive retrogressive processes (chapter 5), as well as a mobilization of those groups and individuals who are able and willing to face up to what must be done. Paramount importance now attaches to the quality of leadership.

Conceptions of welfare and development, their relations and their institutional embodiment likely to be adaptive for the conditions of post-industrialism are radically opposed to the concepts and processes in these areas which have characterized the period of industrialism and helped it forward. For the reasons to which attention has been drawn in this discussion of ideo-existential phenomena formidable difficulties may be anticipated in securing support for programmes based on an assumption of interdependence. The misunderstandings and resistances arise from processes of immense historical depth. The next chapter briefly reviews these.

The Relation of Welfare and Development: Historical and Contemporary Aspects

The Emergence Of a Development Factor in Pre-Industrial Societies
The Medieval Expansion of the Concept of Welfare
The Welfare Function of the State in the Transition to Industrialism
The Development Dynamic of the Protestant Ethic
The Casualty List under Full Industrialism
Forms of Welfare
Contemporary Experience of Growing Inter-dependence
The Direction of Emergent Values

The Emergence of a Development Factor in Pre-Industrial Societies

Settled as distinct from nomadic pre-industrial societies were (and those which survive still are) characterized by an agricultural technology based on the cultivation of crops and the domestication of animals and the many crafts related to these activities. They depended on a favourable habitat and a stable social order which guaranteed the regular performance of the required cycle of activities year by year. However stratified they might become relations between strata were prescribed in binding sets of rights and obligations.

These societies prospered so far as a maximum number of their members were able to contribute—to carry out their tasks according to their station. A functional need was to keep the number of casualities (members unable to contribute for any reason) to a minimum, to restore them to a state of well-being, or, this being impossible, to reduce the extent to which they interfered with what had to be done. This was accomplished largely through the kinship system whose norms operated to limit casualties and to restore or absorb those which were incurred. Everyone was 'in the society', possessing whatever 'welfare rights' were prescribed by the norms–both general norms and those pertaining to particular 'stations', classes or lineages.

These societies were directively correlated with a predominately Type 2 environment (clustered as well as placid), though within the overall environmental domain many areas might retain Type 1 characteristics (e.g. the large number of similar small farms which were quasi-local).

The acquisition and retention of a favourable habitat was all important. The relation of the technology to the habitat rather than the technology itself became salient once the practices and crafts of the technology had established themselves. Their rate of change then became exceedingly slow, for their function was a maintenance one–to maintain the productivity of the habitat, but not to make changes that would overproduce and so exhaust the land.

The group in the society most able to control through land ownership the critical technology habitat boundary became the leading part, acquiring elite status and political power. This frequently became dynastic in form, often combining the sacred and the secular in the temple and the citadel, the twin foci of the city as this arose from the conglomerate of villages comprising the nodes in the network of paths or roads which crossed the clustered agricultural realm (Mumford, 1961). The regulatory power of the leading part could be less coercively maintained internally through the common values and beliefs provided by a single religion congruent with the regular world experienced by most of the society.

The habitat might be endangered by climatic hazard, or some degree of population growth so that its extension became desirable. But extension for these reasons was usually required infrequently and only to a limited extent. By far the greatest danger was the desire of less favourably placed groups to possess for themselves more favourable habitats belonging to others (promised lands). This led to wars; to conquest on the one hand and to the elaboration of defensive power on the other. Defensive could always be turned into offensive power once the appropriate apparatus had been created—a professional army and a superior military technology making use of new types of military formation (such as the phalanx) in control of the leading part. There was therefore a tendency for these societies to expand. The more successful subjugated the less successful. Empires became established within which a 'pax' was maintained which restored the stability required for pursuing the agricultural economy. The disturbed-reactive Type 3 conditions were removed. The pax allowed the component habitats to retain Type 2 characteristics.

Through the expansion process a 'development' factor entered which had far-reaching consequences for welfare. Present in an agrarian empire were large groups not belonging to the original society, but brought under its power by conquest. These out-groups were not accorded welfare rights but were maintained in various states of ill-fare—for the benefit of the imperial power. Extermination was on the whole replaced by exploitation, the additional labour being needed for numerous massive projects such as citadel, wall, temple and pyramid building, quarrying and road making, the digging of irrigation ditches and canals, and the mining of metals for making weapons, utensils and ornaments. The food supply was available; in its 'pacified' and regularly productive state the agricultural habitat could maintain at subsistence level the large numbers of people required without unduly increasing its own more privileged population.

There were two main states of ill-fare which were complementary: being a *slave* and be a *tributary*. The function of slaves and tributaries was to sustain the power of the developing part of the dominant society, which needed it to maintain the pax across the set of habitats comprising the extended domain. Slaves provided labour for household and estate duties in addition to that for large scale projects of the type enumerated. Tributaries provided quotas of food and other staples and taxes—for the controlling elite and the increasingly diversified and talented groups which surrounded them to produce the 'culture of cities'.

These are the conditions which, amongst others, ushered in what Boulding (1966) has called the era of 'civilization', the final version of which seems likely to pass away as the post-industrial society becomes established in relation to the salience of Type 4 environments—when a different set of opportunities and hazards must be encountered. A polarization of states of welfare and ill-fare, of benefits and costs, has characterized civilized societies, however much this polarization may have been reduced in successive forms. This polarization would appear to have been a necessary condition of the advent and persistence of civilization, but the hypothesis may be made that the reverse will be the case in any future form of society likely to survive (cf. Levi-Strauss in chapter 16, pp. 208-9).

The Medieval Expansion of the Concept of Welfare

Successive forms of pre-industrial civilization contributed new 'inputs' to the welfare-development 'mix' which became embodied in key institutions with extensive 'support bases', sometimes co-terminous with the society as a whole. The last major form of pre-industrial society in the West was the feudal society of medieval Christendom. Through universal acceptance of the Catholic Church the beginnings emerged of certain 'ground-rule' equalities which have become part of the Western heritage. The first was *spiritual equality,* the equality of man before God. More particularly in the version of this society which rose in England after the Norman conquest steps were taken on a background of Roman law to advance towards a second ground-rule equality—*legal equality,* the equality of man before the law. In medieval England further steps were also taken through events such as the signing of Magna Carta towards establishing a third ground-rule equality, the *equality of man before the state.* These steps, however incomplete, represent advances over what had been achieved in the Greco-Roman world in the establishment of welfare rights in three socio-cultural dimensions of value.

Medieval society continued ground-rule inequalities, in the economic, social and educational dimensions, as represented in the three estates. Hierarchically organized inequalities persisted side-by-side with newly forming equalities. Yet everyone was re-included in the society under feudalism, serfs, unlike slaves, having defined welfare rights. Medieval society, with all its imperfections, was in its curious blend of values a welfare society in which the pax ecclesiastica replaced the pax romana. Its relevance for present purposes is that it brings out, more clearly than preceding societies in the Western tradition, the idea that welfare can exist in more than one dimension. This idea is likely to be crucial for the transition to post-industrialism especially when to it is added the idea that development can also be multi-dimensional.

The Welfare Function of the State in the Transition to Industrialism

In pre-industrial societies the over-riding concern was with the regulation of stability. Developments occurred as far as they aided this process. It was in relation to development that intervention took place by the political agency of the state or the religious agency of the church or both. The domain of welfare had remained auto-regulative. New phenomena appear when the over-riding concern becomes the regulation of development. The picture begins to reverse itself with remarkable rapidity.

The first new phenomenon is that of the state taking on a welfare function. The welfare state in Britain was not born with the ideas of Beveridge or the legislation which embodied them in the first Labour Government after World War II. It was born in the sixteenth century in the Elizabethan Poor Laws. The enclosure of medieval common land for large scale sheep farming (which the rise of the wool trade had made profitable) created a new class of persons—the unemployed. Renaissance England had to contend with a meta-problem called, in the language of the time, the nuisance of the 'sturdy beggars'—the able-bodied poor. They constituted a new category—the displaced. Such a category had not existed at the height of the Middle Ages. A policy had to be worked out to cope with them as the auto-regulative processes on the feudal desmene which had once taken care of their welfare were no longer available. Either this new class of persons had to be grouped with those, such as criminals or heretics, who were deprived of welfare rights and subjected to punishment (put into ill-fare), or accorded a new type of welfare right which would retain them, in however minimal a way, in the society. The latter was the course taken by the first form of nation-state established during the Tudor period.

The precedent created was that the state intervened as a welfare agent to cope with the social cost of a change process not covered by existing auto-regulative mechanisms. The function of state intervention was to support the developing part of the society, namely that concerned with large-scale production for the market, agricultural in the first instance—but this was the process which was to lead the society towards industrialism. This first action constitutes a recognition of an emergent social process (chapter 3) which had been gathering for something like 200 years. It discloses the nature of state intervention in the welfare area under the conditions of industrialism; to absorb social costs and to keep them separate from the direct costs of market transactions. The effect is to free the market to operate in accordance with the price mechanism.

The Development Dynamic of the Protestant Ethic

Another emergent process which appeared at roughly the same time was the Protestant Ethic. This reversed medieval values concerning welfare. The consequences of this complex and much debated phenomenon for the welfare situation may be summarized (by broadly adapting Weber (1941) and Hagen (1960), as composing the following pattern:

(a) The doctrine of personal responsibility (for salvation) became associated with either that of redemption through good works (rather than by being in a state of grace attainable through the Church); or, in the Calvanist version, that of proving one was among the elect because one's works turned out to be good. Works took on the meaning of work in the sense of gainful employment, the goodness thereof manifesting itself in the prospering thereof. Inward spiritual goodness became symbolized in outward material success. Once the rich man had found it difficult to enter heaven, now it became the poor man.

(b) Material prosperity, because it symbolized spiritual prosperity, could not remain an ancestral right—simply through the inheritance of land. It had to be acquired or demonstrated. Redemption had to be striven for (or election proven). It could be attained only by achievement, acquisition and upwards mobility. This reversed the aristocratic principle and upset the theory of a leisure class. To work counted as a 'good' not a 'bad'; as prestigeful not degrading.

(c) Not to prosper became a sign of wickedness. The poor

could therefore come to symbolize the sinful rather than the innocent and those who remained poor could be perceived as the unrepentant rather than the unfortunate. Calvinistic justice could be seen to be done when those deprived by heaven of spiritual welfare were not allowed to flourish on earth.

(d) These attitudes warranted an ostracism of the poor, their sanctioned shaming, while the poor were expected to internalize these values and feel ashamed of themselves. A stigma attached itself to receiving assistance. To be independent, to help oneself, was sacred. To be dependent, to accept help, was profane. Residues of these attitudes persist at the present time in many groups not only in countries such as the United States which have resisted the welfare state but in Britain which has accepted it.

(e) Poverty could be construed as one's own fault. To bear its misery as justified punishment could be regarded as signifying some remaining virtue if one could not prosper. The undeserving must be uncomplaining; complaint was construed as the aggression of the 'damned'. The meek might be given alms so long as they did not attempt to inherit the earth. Any such attempt tended to be perceived as having a revolutionary rather than a biblical connotation. Revolution, however, was an infeasible strategy for those who were both hungry and unorganized as were the displaced during the first industrial revolution.

Though held intensely and explicitly only by the core members belonging to the more extreme of the non-conformist sects, themselves in any case a minority, in a vague and attenuated form these beliefs coloured the ethos in terms of which the 19th century was to develop. Such values lent support to the developing entrepreneurial part of the society whose welfare (prosperity) had become pivotal. They had the effect of restricting what had to be done for those who were not faring well during a time of troubled change when capital accumulation was a first priority and the initiative of those capable of amassing it required full release. They are the historical root of what pass today as middle-class values, which, however 'square', have released a growth dynamic which threatens to carry us beyond the limits of the regulable. They promoted work addiction and have played their part in producing the work-addicted society from which so many at the present time are seeking to drop out (with the aid of other and equally dangerous drugs).

The Casualty List Under Full Industrialism

Here, using Britain as an example (it being the oldest industrial country), we shall attempt to bring out those aspects of the relations of welfare and development under full industrialism which have led to the confusions which exist at the present time. Development became the leading part of the industrial society when the rate of technological change began to accelerate at the end of the eighteenth century, producing a rise in the rate of economic growth and an associated increase in the size of the population. Studies of the conditions which rendered this possible emphasize the gradual withdrawal during the seventeenth and eighteenth centuries of the state from the regulation of the economy. This is often called the negative theory of government which provided a political philosphy congruent with a free market.

Yet, as the state withdrew from the regulation of development (its function under pre-industrialism) it began little by little to take on the regulation of welfare (its function under industrialism). The Poor Laws of Elizabethan England had been a forerunner as the medieval order dissolved. The controversies over the Spleenham Land Agreement and the Victorian Poor Laws provide later examples. The intervention of the state became necessary because no auto-regulative processes were available to take care of numerous individuals who constituted a new class—the displaced.

One may use a concept of *casualty* to refer to any class of persons unable to contribute, for whatever reason, to the work of a society. Pre-industrial societies recognized three such classes: *the dependent* (the young and the old); *the handicapped* (the initially disadvantaged); *the disabled* (those subsequently disadvantaged whether temporarily or permanently through illness, accidents or wounds). *The displaced* added a fourth class. They were the first form of the unemployed. A fifth class—the *obsolescent* were soon to be added, who in addition to having no further chance of using their skills in their original location could find no place at all to use them. The Luddites were the first instance of the obsolescent, as were the 'sturdy beggars' of the displaced.

The types of people requiring welfare attention increased, as did their numbers, while the capability of the kinship system to cope with them auto-regulatively diminished. The industrialized worker had only his wages as a means of supporting himself and his family once he became a part of the market economy. The unremunerated labour of other members of the family, old or young, and even of the partially disabled or handicapped, could always contribute to the household economy under rural conditions where a certain degree of subsistence farming and cottage industry was preserved. Under urban conditions this could no longer be maintained. The urbanization of the industrial worker broke up the extended family and narrowed the support-base of mutually available kindred. Institutions of some kind had to provide for the welfare of indigent groups, even if only to a minimum standard.

The emergence of the protestant ethic as the dominant ethos of the industrial society created a severe dilemma in the field of values. The belief that it was bad to be poor mitigated against the state developing a welfare function at the rate which the industrial revolution made necessary. Each new reform was won as a reluctant concession only when crisis conditions had been reached. Voluntary agencies, both religious and lay, alleviated the intolerable conditions. The nineteenth century witnessed the creation of the tradition of philanthropy, a feature of the industrial society.

Forms of Welfare

The first welfare measures were *remedial* in the sense that to some extent they removed casualty responsibility from the wage-earner and to some extent restored the casualty to wage-earning capability. Otherwise their aim was no more than *custodial,* whether of the chronic sick, the orphaned or the aged. These measures produced the first of the welfare concepts of the industrial society—that of *social service.* In the latter part of the nineteenth century the state began to take these over from voluntary agencies, though a mix of both remains even at the present time. At first, the state agency was municipal rather than national. National intervention is by and large a twentieth–rather than a nineteenth–century phenomenon, as is the comprehensive range of the services provided.

So far we have considered welfare measures directed towards the individual. The appearance of the urbanized industrial environment necessitated remedial action at the community (ecological) level. Something had to be done about sanitation, infection, etc., the pollution of the environment. Though plagues had periodically visited the cities of an earlier period, they were more tolerable when it was not so necessary to maintain uninterrupted the flow of organized work necessitated by the factory system. The scale and frequency of urbanization was now of another order. To meet such environmental hazards there emerged various schemes of public health. These led to the adoption of *preventive* welfare methods which could only be instituted by a statutory authority, whether at the local or the national level. Since most the country was affected this became the first area in which welfare intervention at the national level developed substance. In the present century preventive measures have not only increased but have tended to replace remedial measures.

The urbanized industrial environment increased the need for public safety. The greater density of the population, its heterogeneity, the large numbers of people in desperate circumstances and the anomie consequent on displacement increased the crime rate, while low wages led to organized protest. An efficient police became necessary. Factories and congested dwellings all burning coal increased the fire hazard. Fire brigades became necessary (they were introduced by insurance companies). The vast scale of drunkenness was incompatible with regular and efficient appearance at work. Licencing regulations were introduced. All these may be termed measures of *social defence,* the second major welfare concept to be elaborated under industrialism. Though directed to the welfare of society as a whole, these measures indirectly benefitted the individual by protecting him against others and himself. Safety regulations were introduced in the factories and an inspection system maintained. At first, measures of social defence were punitive. Gradually, the idea of prevention has spread also to them (an equivalent of planned maintenance).

As time went on a third major welfare concept appeared—*social security.* This entailed providing, through insurance contributions, against the risk of future calamities. One could so lessen the bad consequences to oneself or one's dependents of becoming a casualty. The level of uncertainty was rising as the Type 3 environment gained in salience. Once again, these schemes were begun by voluntary agencies—Friendly Societies. As trade unions arose they also tended to run such schemes. Not until the twentieth century did the state begin to take an active role in social security—with the National Health Insurance Bill of 1911. The first rudimentary old age pensions came only some time later, when unemployment insurance also made an appearance. No subject has exemplified the value confusions in industrial societies more than the controversies centring on unemployment insurance.

The fourth major welfare concept of the industrial society exemplifies the beginning of the interdependence of welfare and development. Activities under this heading *improve* situations and capabilities rather than merely remedying, or even preventing, casualties or hazards. They began at the ecological level through the need to provide an infrastructure of required utilities and amenities for the urban environment. Streets had to be paved. A form of lighting had to be provided. Water supply had to be made conveniently available to factory and home. Public utilities came into existence, many of which were privately financed, but return on capital was low and, as time went on, the state, especially at municipal level, took them over. As social norms regarding required amenities began to rise, parks, playgrounds and swimming baths began to appear and the field of cultural activities to be included. Sports stadiums, concert halls and theatres were built along with libraries and museums.

Welfare activities aiming to bring about an improvement became directed towards the individual as industrial work demanded a minimum standard of literacy. Education began to spread with voluntary agencies such as the church schools paving the way. With the 1870 Education Act, the state intervened and primary education became universally required. Since then the minimum standard has been forced up, with some form of secondary as well as primary education becoming mandatory. Various forms of tertiary education have spread. The notion of improvement has not been restricted to the field of education but has expanded to child care, health and working conditions.

The concept of improvement has been central to the effort of the trade union movement to raise wages and secure a better standard of living for the large majority of the members of the industrial society. State intervention consisted at first in legalizing trade unions (having failed under an earlier ethos to suppress them). More recently, through ministries of labour, it has taken a positive role, helping people to find jobs and providing them with training schemes to increase labour mobility and to raise the level of the work force, now perceived as a major asset.

To sum up. Under industrialism the developing part of the society, concerned with the establishment and maintenance of a free-market economy, becomes auto-regulative. The state withdraws to the margin in this domain but is forced to intervene to a progressively greater extent in the regulation of stability, which is threatened by the dislocations associated with a more rapid change-rate. In consequence, the state acquires a welfare function but can exercise it only against resistance; because, among other reasons, the protestant ethic, which provided the ethos for the leading part of the new society, demands that the individual shall find his own salvation. Welfare measures are introduced piece-meal, in response to crisis, with voluntary agencies paving the way. A remedial approach is followed by a preventive approach and this by an improvement approach; but the earlier approaches persist. Action is taken at the ecological as well as the individual level, both being necessitated by the nature of the urban industrial environment. The domains in which action is taken become ever more numerous, and to concepts of social service and social defence are added concepts of social security and social development. Yet there are few signs of a corresponding tendency for the agencies concerned to co-ordinate their activities, or their outlook. No coherent concept of the welfare function of the state arises to parallel that of the free market, or of the modifications of the free market, even in a country such as Britain which accepts the notion of a 'welfare state'. Yet the underlying need for intervention is unambiguous: to separate social from market costs. Social costs have risen as a direct consequence of the development of the industrial society—just as technological advance has tended to reduce market costs (other things being equal). The predominant value system of the industrial society has perpetuated confusion as to whether, and how far, this trend is a liability or an asset.

Contemporary Experience of Growing Inter-dependence

The situation which has allowed us to see quite clearly that conceptions of welfare and development likely to be adaptive for post-industrialism would be opposed to those which have served the purposes of the industrial society came into existence as the decade of the 1960s took shape. For this last decade has seemed qualitatively different from the earlier post-World War II period in the appreciations offered of key problems arising in the rapidly changing environment. During this period the concept of *development* has acquired a new *trans-economic* meaning in the context first of the 'developing' and then of the 'developed' countries. This has redefined the meaning of *welfare,* which has acquired a *trans-social* connotation—as a contributing resource rather than as a constraining overhead.

Neither of these re-appreciations reduces the importance of considering scarcity factors or makes less arduous the task of deciding priorities in resource allocation. They simply assert that the *best welfare is that which promotes development and that the best development is that which promotes welfare.* Though these new appreciations exist it does not follow that they are universally held or even widely diffused. Very much are they still in dispute.

Before they could exist at all as social beliefs the idea of sustained economic growth had first to attain general credibility, at least in the advanced countries. If economic growth does not proceed in such a country, or if there are interruptions to it, this is not now regarded as a manifestation of a natural law about which nothing can be done, but as something which can and should be prevented. It is now widely though far from universally believed that we have learnt (through the advances made in economics) a great deal about how to regulate economies and that governments should take an active role in the management of economic affairs. This is the reverse of the doctrine which prevailed during the period of industrialism, and which still persists in a variety of business interest groups and upwards mobile 'achievers'.

The new belief began to be established by the experience of the sustained growth of the economies of Western Europe after World War II. Their initial recovery under the stimulus of the Marshall Plan would not in itself have been enough. Additional experience was necessary of the rate and continuity of their progress (despite vicissitudes, which are likely to continue) before this ceased to be regarded as 'miraculous'. In the United States downward swings in the post-war trade cycle have been of short duration; and in recent years the performance of the North American economies has added to the conviction that sustained economic growth will result from informed economic management. Even the difficulties of Britain are seen as an exception that proves the rule; for rather than going backwards, the British economy has simply developed more slowly than others in face of constraints related to the change in its international role and lack of willingness soon enough to accept the implications. Effective international action is more likely to be taken now than in the inter-war period if the threat arises of a serious recession in world trade; moreover, the threat is likely to be detected earlier. Japanese experience has provided further evidence of sustained economic growth, on an entirely different cultural background, and one or two of the developing countries such as Mexico, seem to have 'got over the hump'. Sustained economic growth to varying degrees has also characterized the countries of Eastern Europe in their transition from pre-industrialism to industrialism, even if ideological constraints and lack of management experience have made the going heavy.

Though public ownership is dominant in Eastern European countries, while Western European economies are to varying degrees mixed and North American economies remain to a greater extent private, pattern of ownership is no longer perceived as a critical issue. Rather is weight given to the availability of an immense fund of information and a large repertoire of techniques for managing an economy. Andrew Shonfield (1966) in his book on *Modern Capitalism* has shown how different mixes of public and private power can produce the same results—can be equi-final in systems theory terms. The means may be monetary or fiscal; or more selectively concerned, whether in direct or indirect ways, with the allocation of resources. Usually, all these means are used, however much the agencies may vary, or the patterns of intervention.

The second pre-requisite was to obtain sufficient experience of the conditions under which social development can reinforce rather than constrain economic development for their positive association to become an inherently credible idea in the 'mind' of electorates. That sustained full, or near full, employment is both possible and contributory to economic growth has been demonstrated in Western European countries. There have been gains in social security and the social services of an order not previously regarded as feasible in countries such as France and Germany, which have shown two of the fastest growth rates. Welfare provisions in these countries, though differently patterned from those in Britain and more recent, are as extensive. Greater emphasis is now being placed on improvement strategies as compared with preventive and remedial strategies. Witness the growth of education—not only initial but life-long education. In higher and continuing education the most spectacular developments have occurred in the United States, and more recently in Canada. No longer is it merely an idea but an experience that re-training over and above initial training is becoming a regular part of life, and that a person may now expect to have two or three careers instead of one. That social costs need not prevent economic growth but may be a condition of its continuation by the benefits created is becoming part of the new system of belief (however difficult the measurement of benefits may prove to be and though resistance to this belief by its industrial predecessor remains strong).

Paradoxically, problems of priorities in resource allocation and of multi-value criteria in making choices are perceived as more difficult in consequence. Only in relation to the transition to post-industrialism are the immense scale of social costs and the problematic nature of associated benefits beginning to be confronted. New appreciations in this area make a designation of affluence look premature even when applied to the most advanced countries. In these there may be affluent people but in none does a general condition of social affluence characterize the texture of the society as a whole. Confusions about and conflicting attitudes to taxation pin-point the dilemmas involved in mobilizing resources to meet rising social costs. It is symptomatic that a number of advanced countries are being forced to consider radical revisions of their taxation systems and that schemes brought forward after extensive inquiry should arouse widespread controversy.[1] Problems of inflation control and incomes policy remain unresolved.

Nevertheless, in some countries, particularly those of Scandinavia, a point has been reached where welfare rights in all six of the dimensions listed in chapter 10 are perceived as belonging to generally acceptable and operating social norms. Welfare rights are thought of as also constituting development rights. Emphasis is placed not so much on minimum standards as on the realization of potential, for the sake of the individual and for society. The belief is gaining ground in most advanced countries that a society which develops and utilizes to the *full all* its *human resources* may be attainable in a *foreseeable* future rather than in some unforeseeable millenium. So far as it appears to critical classes of persons such as the highly educated young or severely disadvantaged minorities that no evident headway is being made towards this goal, we may expect increasing impatience and rising protest, widespread alienation and deepening dissociation. It is an open secret that the technologies required to remove ill-fare are already available. The barriers are in our attitudes and values which remain confused in many sections of the population between those appropriate to industrialism and those which foreshadow post-industrialism. The turmoil over rival 'counter-cultures' and the withdrawal of status respect (Hagen, 1960) from 'the establishment' bear witness the world over.

The Direction of Emergent Values

These confusions and their attendant conflicts are acute in the United States and Canada—in the former, more than in the latter, in view of its size, complexity, and the character of some

[1] c.f. Jacoby, Neil H., 1967, Canada's Tax Structure and Economic Goals, A Critique of the Royal Commissions on Taxation, York University, Faculty of Administrative Studies.

of its traditions concerning individualism and competition. Both these countries are moving towards post-industrialism faster than Western European countries, but in so doing are entering crisis conditions in the confrontation of their under-developed parts. The extent, magnitude and heterogeneous nature of these parts are only now in the process being fully disclosed. The qualifications (of nature, nurture and opportunity) required for entry into the developing parts are rising. There is small chance of policies which will meet the gathering dangers being either formulated (or accepted) or of adaptive institutions being built to implement them effectively (or to evaluate them convincingly) until present confusions over obsolete meanings and relations of welfare and development are replaced by new appreciations which express their inter-dependence and complementarity.

The required response lies, as suggested by Frank (op. cit) in planned collaborative attacks directed to clear purposes and leading to concerted action–if the emerging meta-problems are to be solved: 'This implies the need for an over-all, comprehensive policy that will assert the criteria for choices and decisions. With a clear statement of policy, those who make social decisions can be guided, as if by "an unseen hand", when exercising their autonomy to integrate their efforts by collaborating with others who are responsive to these same criteria. Without a statement of basic criteria for national policies, the various specialized programmes and the separately located authority of governments and private agencies will continue to plan and execute their separate and often irreconcilable programs.'

The basic criteria for which he asks will become available so far as a new set of values emerges which match the needs of the oncoming post-industrial society, much as the Protestant Ethic matched those of industrialism. The direction in which these new values lie has in the writer's view already been sufficiently discerned to be capable of general statement. An attempt at such a general statement is made in the accompanying table. This seeks to relate

(1) the cultural values which inform the choices of the individual as a member of the extended informal system (social aggregate) of his society to

(2) the organizational philosophies which regulate the

Table 1 Changes in Emphasis of Social Patterns in the Transition to Post-Industrialism

Type	*From*	*Towards*
Cultural values	achievement	self-actualization
	self-control	self-expression
	independence	inter-dependence
	endurance of distress	capacity for joy
Organizational philosophies	mechanistic forms	organic forms
	competitive relations	collaborative relations
	separate objectives	linked objectives
	own resources regarded as owned absolutely	own resources regarded also as society's resources
Ecological strategies	responsive to crisis	anticipative of crisis
	specific measures	comprehensive measures
	requiring consent	requiring participation
	damping conflict	confronting conflict
	short planning horizon	long planning horizon
	detailed central control	generalized central control
	small local government units	enlarged local government units
	standardized administration	innovative administration
	separate services	co-ordinated services

Notes: The terms used are intended to be self-explanatory but reference may be made to McClelland (1961) on achievement; Maslow (1954, 1967) on self-actualization; Tomkins (1964) on the regulation of negative affects (such as distress) and positive affects (such as joy). The need to regard corporate resources as belonging to society as well as the corporation became a major theme in *A Statement of Company Objectives and Management Philosophy* (Shell Refining Co., London 1966), elaborated by Paul Hill in *Towards a Management Philosophy* (Nigel Farrow, London 1971).

behaviour of the institutionalized groups to which he belongs and

(3) the ecological strategies followed by his government (at whatever level).

For each domain patterns persisting from industrialism are contrasted with those emerging in relation to post-industrialism. One will not give place entirely to the other but the established will become redefined in the context of the emergent (c.f. chapters 6 and 14).

A stronger emergence and wider diffusion of such patterns is a necessary condition for a new political philosophy of the type called for by Frank to develop and to gain acceptance. Only when a new basis for politics is so provided will the sufficient conditions be established for accomplishing the transition to post-industrialism. The accelerating change-rate has created new levels of environmental complexity and uncertainty while its unevenness produces manifold and increasing social 'fragmentation, segmentation and dissociation' (chapter 5). These trends endanger our adaptive capability. To meet them our only hope is to take the active rather than the passive role. Only so far as we learn to take the active role will our capability increase to deal with meta-problems and to regulate social processes at the ecological level. This will entail building a service state on principles which recognize the inter-dependence of welfare and development. By promoting both in the primary units of society we can give our world a chance (probably its last) to regulate itself.

Chapter 12

The Structural Presence of the Post-industrial Society[1]

The Drift Towards Post-Industrialism
The Basis of the Structural Model
Changes in the Socio-Technical Power Base
Changes in the Economy
Changes in Occupation and Education
Changes in Leisure and Unemployment
Changes in the Family
Changes in the Environmental Context

The Drift Towards Post-Industrialism

It is an honour that you should aske me to offer this address. More importantly, that you are prepared to give a hearing to someone finding his way to an interest in the planning process from a background in social psychology is a sign that you—who are the leading group in urban planning in your country—at the action as well as the academic station—are among those seeking to extend the inter-disciplinary range of environmental studies. This gives me a feeling of scientific and professional reinforcement, which all who 'engage' with the problems of the contemporary environment could do with more often.

Let me now state my premise:—that an irreversible change process is proceeding in the world, at an accelerating rate but

[1] This chapter and chapters 13, 14 and 16 are expanded from the Keynote Address given at the Annual Conference of the Canadian Institute of Town Planners, Minaki, Ontario, 1968. They retain the form of an address. A revised version was presented at the Joint Conference of the International Association for Social Psychiatry and the World Federation of Mental Health, Edinburgh, 1969. This was circulated for discussion at the Conference on 'Organizational Frontiers and Human Values', Graduate School of Business Administration and co-sponsored by the National Training Laboratories for Group Development, U.C.L.A., 1970. Much new material has been added as a result.

with extreme unevenness, both within and between countries, which I shall refer to as a drift towards the post-industrial society. The advantage of the term 'post-industrial' lies in the implication of its metaphor: that we may not assume that the present social order will continue indefinitely; rather must we prepare ourselves to assist the emergence of a society radically different from the industrial societies which have evolved in the last two hundred years—whether these remain substantially capitalistic or have taken on either a mixed or a socialistic complexion.

I have used the word 'drift' to indicate that the process is not under control. In *Value Systems and Social Process,* Sir Geoffrey Vickers (1968) warns us *as a species* that unless we succeed in establishing a control within some acceptable limit the danger is extreme that critical aspects of our environment and our lives will become unregulable. For we now have 'species problems' which we ourselves have created, and species problems are different from national or international or economic or social problems raising as they do questions of our bio-social survival as humans. To sense them changes for the perceiver the character of the 'world landscape'. During World War I Kurt Lewin noticed that he could not escape from the pervasive presence of what he called the war landscape. Phenomenologically, no less than militarily or politically, the landscape of peace could not be experienced. Today, on the basis of the kind of 'information' I now receive I find I cannot re-capture that sense of solid earth I once took for granted.

To understand why this is so requires an appreciation of what has become the salient characteristic of the contemporary environment, namely, that it is taking on the quality of a turbulent field (chapter 4). Forms of adaptation, both personal and organizational, developed to meet a simpler type of environment no longer suffice to meet the higher levels of complexity now coming into existence (chapter 6).

Let me use this line of thought to pin-point the planner's dilemma: *the greater the degree of change, the greater the need for planning, otherwise precedents of the past could guide the future; but the greater the degree of uncertainty, the greater the likelihood that plans right today will be wrong tomorrow.* Such dilemmas produce 'ecological traps' (Vickers op. cit.).

The analysis now to be offered represents an attempt by

looking for certain characteristics of the future in the texture of the present to search for ways of getting out of the trap. For just as the past can persist in the present so can the future lie concealed within it (chapter 3). My endeavour will be to show that in the most advanced countries the post-industrial society is, structurally, already present. Far from being an imaginary entity which may possibly happen in the year 2,000 in a structural sense it has already 'occurred'. In fact it has been building up for some considerable time and its outline form, can we but see it, is present in our midst. This will be my first theme. By contrast, what has not yet occurred, and what is not building up at the pace required, is any corresponding change in our cultural values, organizational philosophies or ecological strategies. This will be my second theme (chapters 13 and 14). The mismatch which this creates is likely to prove our stumbling block, preventing us from accepting 'the burden of choice' (chapter 15). Until we recognize that we are not still industrialists, we cannot develop the capability, though we have the resources, to shape our future to good advantage. To turn the mismatch into a good match constitutes the challenge before all of us. A central issue which this raises concerns the establishment of a new relation between planning and the political process. This will be my final theme (chapter 16).

The Basis of the Structural Model

In looking at structure I have depended extensively on work carried out by Bertram Gross (1968a) under the sponsorship of the Twentieth Century Fund–'An Overview of Change in America'–as regards a good deal of the data and certain of the ideas–not that he is responsible for the particular method used or the model constructed.

For convenience the model is presented in a set of descriptive tables. The calculations on which the various sections of Table 1 are based make use of Kendrick's (1967) and Schultz's (1961) revisions of national economic accounts and of Gross's further modifications of these; they also incorporate his (1966) and others' work on social accounts. Though reference is to the U.S., the model is applicable to any of the advanced countries, which differ only in degree and rate of change in the respects considered. Canada is closest to the U.S. in most of these, though not in some of the other features which would be included in a more comprehensive model.

The method consists of selecting 21 critical aspects of society which have undergone a 'phase change' from an industrial to a post-industrial pattern in the thirty years which separate the mid or late '30s from the mid or late '60s, so that changes in the last thirty years may serve as a base for considering—though not predicting—changes in the next.

The phases are grouped in six domains. The first identifies the 'leading part' and its 'characteristic generating functions' (c.f. chapter 2)—science and the technologically advanced industries together with the new power acquired by the elites involved. This leading part has change-generating effects on the other five domains—the economy, the occupational and educational system, leisure and unemployment, the family, and the wider environmental context. But these effects are two-way. Each domain has relations with the other domains as well as with the leading part, simultaneous or successive, so that the set as a whole comprises a social field.

Vision thirty years backward and thirty years forward has an existential property. It is as far back as a man in his prime may look without being confused by recalling his adolescence and as far forward as he may look without becoming confused by anticipating his senility. This is not without relevance to the longest time-span which has psychological reality as a planning horizon. Conceptually, this time-span corresponds to the 'back—and forward—reference periods' of medium-term directive correlations as Sommerhoff (1950) has developed this concept in his theory of adaptation. Medium-term directive correlations represent ontogentic time—and ontogentic time may perhaps be the best type of time for a planner to keep if his conjectures are to be living.

In Table 1 the first column lists the aspects selected; the second, patterns salient thirty years ago; the third, patterns salient, or becoming salient, 'now'. The result is a model consisting of a set of contrasting patterns which exhibit the scope and nature of fundamental and irreversible changes which have led to the structural, though not the cultural, presence of the post-industrial society.

Changes in the Socio-Technical Power Base

The first group of aspects (1-3), Table 1(a), represents the socio-technical power-base of change, the configuration of factors which show that a phase change has occurred in the character of the change-generating forces. As Daniel Bell (1967) has emphasized, 'what has become decisive for society is the new centrality of theoretical (scientific) knowledge'. This, as distinct from knowledge derived rather than tested empirically, 'has become the matrix of innovation'.

Table 1 Patterns of Comparative Salience 1935-1965
(based on U.S. data)
1(a) The Socio-Technical Power Base of Change

Aspect	*Salient '35*	*Salient '65*
1. Type of scientific knowledge	Empirical	Theoretical
2. Type of technology	Energy → Assembly line	Information → Systems management
3. Politically most influential advisers	Financiers and industrialists	Scientists and professionals

This change is related to the emergence of a new technology based on a new concept–information. The advent of the computer constitutes a step-function advance in human ability to handle complexity without which the post-industrial society would be unrealizable. Factor increase in computer capacity since World War II is a graph with a vertical climb (Ayres, 1966). Fusion of the information and energy technologies has produced systems engineering. The systems aspect was soon appreciated as general. This is building a new capability to understand large-scale systems in a variety of fields and post-industrial society is characterized by an increase in their number which would otherwise be overwhelming. Systems analysis does not in itself, as some would suppose, solve the problem of multi-valued choice. It does, if we so will, enable us to confront it. This confrontation is becoming a central issue as post-industrialism emerges.

These changes have occasioned a third: the replacement of the financial-industrial elite by the professional-scientific elite as the most influential advisers in the corridors of power (though they still remain uncommon in the seats of power—to Lord Snow's regret (1961)). The financial-industrial group is likely to continue as a powerful 'third force' to gear the mobility of capital to the science-inspired rate of change in productive possibilities, as in the emergence of conglomerates. Governments, however, at least in the advanced countries, are now themselves the wealthiest capitalists, controlling the markets for the most advanced industries. The resources they most need are those provided by knowledge-makers and knowledge-appliers. These inter-dependent fraternities own the means of their own production—which are in their heads.

Changes in the Economy

The second group of aspects (4-7), Table 1(b), identify major changes which have taken place in the character of the economy. When goods and goods-related services are separated from service-related services and person-related services, these latter now account for more than half the GNP, for more than half of total employment, and for most of the gains in numbers employed. If the activities of all self-employed professionals, private non-profit and non-commercial organizations in

1(b) The Structure of the Economy

Aspect	*Salient '35*	*Salient '65*
4. Contribution to GNP	Goods and goods related services	Service and person related services
5. Sector	Market	Non-market
6. Leading private enterprises	Domestically centred	Internationally centred
7. Costs	Marketable commodities	Supporting social and urban environment

whatever field (health, education, welfare, research, philanthropy, religion, community, conservation, etc.) are included along with those of government agencies and services (at all levels) they now contribute more than half of what the market sector contributes. If, however, a value is added (following Kendrick and Gross) for households–non-paid personnel, capital, productivity gain and volunteers–the contribution of the total non-market sector begins to approach (some would say exceed) that of the market sector. The relevance of pooling all components of the non-market sector is that it releases us from being controlled by the industrial 'image', which assumes that the market sector alone counts in producing the wealth of a society.

Within the market sector the larger enterprises, which account for the larger part of the activities (Galbraith, 1967), have in scope become international. Multi-national corporations–of whom a few hundred have reached giant size–have emerged as a major phenomenon of the post-industrial society. Their countervailing power to that of nations arises from the extent to which they are bringing into being a *global industrial estate* (Perlmutter, 1969). This is a different concept from that of world trade conducted by large numbers of firms manufacturing mainly in one country. What may be meant by a national economy needs review. Politically, the emergent requirements of multi-national corporations are creating the rudiments of what Perlmutter calls trans-ideological space even in East-West relations. Failure to recognize the claims of host countries encourages nationalization so that they must negotiate more modest 'take out' profits and contribute to indigenous growth. Moreover, their growing need is not only to employ in their operating establishments the nationals of the many countries in which they do business; but, in order to gain understanding of their multi-cultural environment, to include the best talent from any of these countries on their top boards. They are passing in Perlmutter's view (1965) from an ethnocentric, through a polycentric stage from which they are taking first steps towards a geocentric stage. Some of their members are developing what may be called a transnational identity, as is observable in international civil servants. Maybe everyone now needs a transnational component in his identity.

The world youth culture is moving in this direction—on however different a premise from the multinational corporation—but extremes are supposed, at least sometimes, to meet.

Considerable discretionary purchasing power is now enjoyed by the many as compared with the few, but the cost of bringing into being and maintaining in a viable state the total environment (physical and social, urban and rural) required to support a high level of personal consumption is approaching or (again some would say) is beginning to exceed the cost of producing the commodities themselves. Though, through technological change, mass consumption societies produce more marketable commodities more cheaply, in so doing they are generating social costs whose scale is only now beginning to be realized. Social costs must be met if the transition to post-industrialism is to be made without increasing disorganization. The relation of market to social costs is the area about which present confusion and misunderstanding are greatest, controversy most bitter and conflict most difficult to resolve—since it is an area where questions of basic values cannot be avoided. It is also the area of most direct relevance to those concerned with improving 'the urban condition' (Duhl, 1963), who find themselves constrained by these value differences, while facing the challenge of producing new forms of 'the built environment' (c.f. Holford, 1965) which will increase rather than diminish 'the quality of life'.

Post-industrial society may be seen as a service society in which the market, however important it may remain, is becoming sub-dominant. Economically, it will be a more international society, though one in which nationalism and 'sub-nationalism' will continue to assert themselves. Though remaining a mass-consumption society, with social costs beginning to overtake market costs, it will tend to become, to use Daniel Bell's term a more 'communal society' in which the individualistic values of industrialism will be less pronounced. Not that these values, any more than the other industrial features listed, are likely to diminish beyond a certain limit. Rather is the question one of change of emphasis. This different emphasis does, however, imply a change of structure. A new type of social balance must somehow be struck if the new order of complexity is to be regulated (chapter 13).

Changes in Occupation and Education

The third group of aspects (8-11), Table 1(c), is concerned with phase changes in the occupational structure in relation to employment and education. The large mass of jobs in manufacturing industry depending on unskilled, semi-skilled labour and even skilled labour when this is manipulative, is

1(c) Occupational Structure and Education

Aspect	*Salient '35*	*Salient '65*
8. Composition of work force	Blue collar	White collar
9. Educational level	Not completing High School	Completing High School
10. Work learning ratio	Work force	Learning force
11. Type of career	Single	Serial

being automated and computerized. The jobs that remain, or are being created, involve perceptual and conceptual skills on the one hand and interpersonal skills on the other. White collar already outnumber blue collar workers. Were it not for the many new types of activity at the sub-professional level which automation has created, and the not unrelated growth of the service industries, there would already be signs of the unmanageable increase in unemployment which was widely predicted as a consequence of automation only a few years ago.

A first consequence of increasing automation is that the qualifications required to enter 'the world of work' are being raised to a new level. This gives one meaning to the fact that 75% of the present school generation is now completing high school. A survey of Science Policies of the U.S.A. prepared for UNESCO (1968) by the National Science Foundation states that 'nearly half of this total group (about one-third of the females and half of the males) enter institutions of higher education. Of this proportion, a little more than half receive a baccalaureate degree. In short, more than 15% of the youth (about 25% of the males) in the United States now obtain a higher degree.' Thirty years ago less than 50% completed high school, less than 12% started college, less than 6% finished. Completion of grade rather than high school expressed the educational norm for the bulk of the population of the industrial society.

A second consequence relates to the maintenance effort now required to remain in the work force, given the obsolescence-rate of skill and knowledge. Pertinent here is a new finding from Gross' work—that the 'learning force' is already greater than the work force. When all those in some form of continuing as well as formative education are put together, they outnumber those working at their jobs. This was expected in the U.S. by 1975 but improved statistics showed it had already happened in 1965.

A linked change is the number now proceeding from a single to a serial career—based on a growing realization that an initial occupation is unlikely to last out a working life. This change is already salient in the generations on the younger side of their mid-careers.

Changes in Leisure and Unemployment

The next two aspects (12-13), Table 1(d), consider leisure, first in relation to employment, then to unemployment. The ratio of leisure to working hours (among waking hours) has become positive for the bulk of the employed population if the year (including weekends and holidays) is taken as a whole. A four day week is on trial in a number of places; 'time off in lieu' is frequently preferred as the method of compensation for overtime in industries where the basic wage level is high. While stability of income is sought through long-range employment contracts, protection of leisure is sought by limiting the work week. Trouble arises when the level of income now expected cannot be obtained without encroachment on the amount of leisure now also expected. This is what the expectation of and demand for 'affluence' in the mass of the population currently means. The only groups for whom the work-leisure ratio remains negative are the members of the scientific and professional elites and their executive and political counterparts who together constitute the 'meritocracy' (Young, 1958). For these the distinction between work and leisure is frequently

1(d) Leisure and Unemployment

Aspect	*Salient '35*	*Salient '65*
12. Work/leisure ratio	Working hours	Leisure hours
13. Character of unemployment	Cyclical though large	Permanent in disadvantaged

unreal (or alleged to be)—though the cardio-vascular system may not agree.

If the development of opportunities for effective use of more leisure in the employed population constitutes one challenge to innovative urban planning, a more immediately severe challenge arises from the need to ameliorate the circumstances of those whose 'leisure' derives from their unemployment. In the late '30s the unemployed were perceived as a manifestation of incomplete recovery from economic depression; if Keynes could be used to manage the economy they would all but disappear, and post-war Western European experience for two decades seemed to confirm this expectation. For some time, however, another process has manifested itself in the U.S., and in Canada, and is beginning to affect certain European countries. This has brought about the permanent presence, at the bottom of—or rather outside—the main society, of a large, heterogeneous and growing class of disadvantaged and disturbed individuals unequipped to meet the higher requirements now obtaining as a condition of entering, or maintaining a position in, the changed occupational structure. They become 'ecologically trapped' in poverty. When they obtain work this is apt to be of a kind so poorly paid that, if they have large families, they remain below subsistence level.

In the U.S. that the negro should be the core member of this class has produced a situation of deepening crisis, which suggests that a limit of the Vickers type has been passed. The nature of the underlying problem and its growing scale are, however, general and a function of the presence of post-industrial features in a society continuing as if still only industrial. This type of problem has no solution in terms of the values of industrialism. New understandings are required such as that of the relation of welfare and development discussed in chapter 10. As was noted there and in chapter 7, Michel Chevalier has called many-sided problems of this type, which require new appreciations in the transition to post-industrialism, 'meta-problems'—his central point being that not only have the problems themselves developed wide ramifications but that their diffuse extension is also becoming widely perceived and attitudinally elaborated in emotionally laden psycho-social belief systems.

The urban planner, it would follow, must learn to work in the meta-problem rather than simply in the problem frame of reference; and to take a leadership role in persuading others to act in terms of the underlying rather than the superficial issues. A number of people at the 'grass roots' have been found in several studies, including one by the Tavistock Institute of communications in the National Farmers' Union of England and Wales (Higgin, Emery & Trist, 1965), to understand this better than many of those in decision-maker roles who tend to cling to whatever part of the problem lies within the jurisdiction of their departments. Our institutional structure has been built up in a compartmentalized way to suit the conditions of industrialism.

Changes in the Family

The fifth group of aspects (14-16), Table 1(e), concerns the family, where phase changes producing considerable disturbance have also occurred. The nuclear family is being complemented by the 'semi-extended' family (Gross, 1968b), characterized by the presence of more than two generations (three commonly, four and even five not uncommonly—in communication, thanks

1(e) Family Structure

Aspect	*Salient '35*	*Salient '65*
14. Basic type	Nuclear	Semi-extended
15. Inter-generational conflict	Local	Societal
16. Hard goods investment	Businesses	Households

to the technology now available, though not necessarily under the same roof). This change is occurring at a time when inter-generational conflict is rising directly as a result of the faster overall change-rate, which gears each generation to a different future. With the current rate of change it may be desirable to think in terms of sociological age-cohorts, age groups with characteristically different life experience and not just biological generations. If the immediate past is any guide we may have to think in terms of two or more sociological age-cohorts per generation.

Meanwhile, the prolongation of the educational process is increasing the economic dependence on their parents of adolescents and young adults (now marrying at an early age) at a time when these same parents are having to take more responsibility for their own elders. The ensuing strains are both emotional and economic. They have not a little to do with student unrest. But the economic and the emotional aggravate each other particularly in the lower middle class–the post-industrial version of the industrial 'working class'. This accounts in no small measure for their hostility to disadvantaged groups who threaten them as a tax burden, as devaluers of their property, as lowerers of the quality of education available in their neighbourhood, or as challengers of their values and way of life. The 'hard hat' has emerged in the U.S. as the symbol of this social segment. Hard hats and their families may constitute a segment of 40 million but they do not constitute a majority (and are no longer silent). No one group constitutes a majority in a pluralistic society. In Britain their equivalent became referred to during the '50s as the rich working class. Having tasted middle class standards through high piece-rate earnings on overtime, this class will not be denied their continuous possession which they see others more securely enjoying. Wage increases of a new order of magnitude have recently been demanded, and acceeded to, in a number of countries. 'Relative deprivation' now means deprivation from middle class standards. A sociological theory of distributive justice such as that of Homans (1958) needs to be taken into account by economists seeking to explain inflation and to prescribe remedies for it.

The more established a middle class family becomes, the more highly does it capitalize itself. More hard goods investment now goes into homes than into business enterprises in the U.S. (Gross op. cit.). This is a result of complex forces, intensified by the cost of purchasing personal services. If one effect is to increase the isolation of the middle class family from the surrounding society, another is to create innumerable pockets of under-employed resources which make the middle-class suburb the most cultivated economic wasteland in post-industrial society. In a string of imprisoning residential areas such as Westwood, Beverly Hills, Bel Air, Brentwood and Pacific Palisades in the Los Angeles basin one encounters the social form which the desert has now taken in California. For the urban planner it creates problems equalled only by those, however different, of the slumming up of the inner city. This is the American form of the problem. In the developing countries the outskirts are the parts of the city which are slumming up. In Europe the great cities still retain life in their hearts but their vast middle suburbs are dead.

In a sense we have all become 'ghettoized'. This image targets the extent to which social defences of superficiality, segmentation and dissociation (chapter 5) have been mobilized by the societal drift towards post-industrialism. A different type of more integrated urban environment is a necessary condition for correcting this—one in which daily contact can take place between members of a wide cross-section of a society. Without the informal and continuous 'reality testing' which this permits, hostile fantasies may be expected to develop which will entrench prejudice between the various sections of an 'apartheid society'. Newcomb (1947) called this phenomenon autistic hostility.

Changes in the Environmental Context

My last group of aspects (17-21), Table 1(f), concerns certain broad features of the environmental context, beginning with the organizational. The phase change here is that the salience of the large single organization, such as General Motors, has been replaced by the salience of large inter-organizational clusters, such as the NASA-space complex. There are wider and more vaguely bounded entities—people as different as Lyle Spencer, head of an IBM subsidiary (*vide* Gross, 1968b), and a social critic such as Michael Harrington (1965), now talk about the social-industrial complex, rising in the wake of the military-industrial complex, the first being a response to the urban crisis as the second was to the cold war. There is talk also of a medical-industrial complex and an educational-industrial complex—even a musical-industrial complex created by the sales success of the 'pop' counter-culture (the youngest millionaires wear long hair). These clusters represent complex inter-relationships between government agencies, industrial enterprises, professions, universities and, not unimportantly, some of the sub-cultures based on age-grading. The clustering is becoming increasingly pluralistic.

The inter-organizational cluster has been a principal cause of the second change in environmental context, the appearance of the inter-metropolitan cluster. Whether or not we care for any of its particular manifestations to date, the inter-metropolitan cluster would seem to be the urban intimation of the post-industrial society. It is many years since Sir Patrick Geddes (1968) introduced the concept of the con-urbation to denote the advance of the city towards megalopolis. His pupil Lewis Mumford (1961) believes that *urbs* as *civitas* is a vanished form.

1(f) The Environmental Context

Aspect	*Salient '35*	*Salient '65*
17. Organizational	Large single organizations	Inter-organizational clusters
18. Urban	Single metropolitan areas	Inter-metropolitan clusters
19. Rural	Quasi-autonomous	Urban-linked or dissociated
20. Pollution	Within safety limit	Passing safety limit
21. Natural resources	Treated as inexhaustible	Feared as exhaustible

A regional alternative which might permit an organic transformation of complexity in diversity has been proposed by Friedman and Miller (1965) in their concept of the 'urban field'.

Those parts of the rural environment (such as 'agro-bis' farms) which are economically viable have become part of urbanity. Those (such as outlier homesteads) which are not have become dissociated. They decay. Their inhabitants are outside the main society as much as those of the city's slums. They are as difficult to reach with any effectiveness (Abramson, 1968). The sub-economic farms still scattered across the breadth of Canada could flourish, so long as the mix of Type 1 and Type 2 conditions which produced them remained salient in their environment; but they are perishing in the mix of Type 3 and Type 4 conditions which now prevail. The Federal-Provincial Agreements reached by the Agricultural Development and Rehabilitation Administration to assist such backward areas as N.E. New Brunswick, the Gaspé and the Interlake Region of Manitoba represent one of the most sophisticated attempts (whatever the vicissitudes encountered) by a modern government to meet such problems by involving the local populations.

The present level of environmental pollution cannot continue without inflicting irreparable damage and the rate of consumption of natural resources cannot continue as if the supply were infinite. Though realization of all this has spread with remarkable rapidity in the last four or five years little has been done about any of the big items. These require agreements between nations as well as between interest groups within nations.

The structural presence of the post-industrial society discloses that we have reached or will soon reach a number of limits critical for our survival. Unless we can learn new methods of social regulation, the chances are small of our beginning to realize the immense possibilities that now exist for improving the quality of life–and the risks extreme that there will be a number of large scale disasters, some of them imperceptible for quite a time because of their slow tempo.

Chapter 13

The Cultural Absence of the Post-industrial Society

A Three Dimensional Cultural Model
Self-Actualization
Self-Expression
Inter-dependence
Capacity for Joy

A Three Dimensional Cultural Model

This brings me to the second theme, the absence of a culture congruent with the needs of the post-industrial society despite the fact that post-industrialism is structurally very much in evidence. Let me ask what the salient cultural patterns are today compared with thirty years ago, more especially those related to our core values, whether personal, organizational or political. The answer can only be that they are largely the same. Is it surprising therefore that we are witnessing a mounting crisis of alienation whose manifestations increase in variety and in intensity, whether expressed as withdrawal or protest? Maladaptive trends are only too plain in many of the more extreme of the contemporary sub-cultures which represent short-circuiting strategies, self-limiting attempts at complexity-reduction. This is not to say that adaptive trends are nowhere to be discerned, but that they are nearer the horizon than the main sky.

If I ask what kind of new values can be regarded as appropriate, my answer can only be in terms of the following criterion: that they must be values which enhance our capability to cope with the increased levels of complexity, inter-dependence and uncertainty that characterize the turbulent contemporary environment. Evidence is mounting that the individual by himself, or indeed the organization and even the polity by itself, cannot meet the demands of these more complex environments. A greater pooling of resources is required; more sharing and more trust. Appropriate emergent values will need to be congruent with Emery's second design principle—the redundancy of functions. They may be expected to be communal rather than individualistic regarding access to amenities, co-operative rather than competitive regarding the use of scarce resources; yet personal rather than conforming regarding life styles and goals.

As regards access to scarce amenities and resources we may expect fewer exclusively private parks and art collections; less hoarding of scientific and technological discoveries as potential trade secrets. There will be a greater sharing of the more expensive items of equipment, the greater benefit secured to all being more highly valued than unique possession. There will be more leasing and renting; less owning. Involved is more than the philanthropic (industrial) concept of lending to the disadvantaged. The underlying issue is that of greater utilization. There will be greater tolerance for different life styles—in other groups and in phases of one's own life. This, hopefully, will lead to some reduction of one-dimensionality.

The direction of new values will be opposite to that which value-formation has taken in industrial societies, moulded as these have been by the Protestant Ethic. One may, therefore, expect resistance to their establishment and diffusion. Such resistance, which is both conscious and unconscious, is responsible for a good deal of the cultural lag which has prevented more extensive co-operative action from having already been taken concerning problems of the urban environment (chapter 10).

Tables 2(a), (b) and (c) represent a model of 17 key social patterns persisting from industrialism (column one). Contrasted is a set of recently emergent patterns (column two) which, though their hold is precarious, may be regarded as congruent with post-industrialism and therefore likely to gain in salience. The table as a whole was presented in chapter 11; the sections are re-presented here and in chapter 14 at the points where they are discussed. They are grouped in three domains: cultural values carried by the individual as a member of the social aggregate; organizational philosophies embodied in the practices of the formal organizations to which he belongs in various task environments; ecological strategies through which governments and interest groups (at any level) seek to regulate the contextual environment of his society. Events and processes in the three domains influence each other: if in the same direction, a self-consistent and coherent ethos may be expected to arise; if contradictory, the maladaptive defences described by Emery in chapter 5 may be expected to increase.

The model may be compared with that for the structural variables presented in the various sections of Table 1. It may be related also to the four dimensions of ideo-existential processes discussed in chapter 10 where some of the problems were touched on which arise from variations in the interpretation, degree of acceptance and rate of diffusion of new values in the different parts of complex societies. Though the emergent patterns to be described may all be said to exist as cognitive orientations, as yet they function as motivators of behaviour only for relatively small numbers of people and are embodied in the practices of only a few organizations. They are perceived in widely differing ways in different groups. Our concern is with the extent to which such new patterns, as patterns, have properties which are likely to aid adaptation to complex environments. Methods for the empirical examination of their actual state in the social present and their probable state at various points in the social future were discussed by Emery in chapters 2 and 3.

Table 2 Changes in Emphasis of Social Patterns in the Transition to Post-Industrialism
2(a) Cultural Values

1. Achievement	Self-actualization
2. Self-control	Self-expression
3. Independence	Inter-dependence
4. Endurance of distress	Capacity for joy

Under 'cultural values' are listed four corner-stones of industrial morality: achievement, self-control, independence, and endurance of distress (grinning and bearing it). But while psychologists are busy giving tests to discover how 'need achievement' is progressing in developing countries, planners concerned with the advanced countries might be better advised to design the 'habitats' of the future from the post-industrial list. This would include self-actualization, self-expression, inter-dependence and a capacity for joy. The new list will not replace the old but will gain in relevance and centrality. It will produce a re-structuring. The new items, like the old, are inter-related among themselves. As a set they will come out differently through emerging on the background of the prevailing items. Post-industrial values represent retrievals, rediscoveries of pre-industrial values which have been diminished, or lost, in industrial society. The gestalt which includes both will be new, for the industrial set had scarcely emerged in the pre-industrial period.

Self-Actualization

Self-actualization is a concept associated with the need hierarchy introduced by Abraham Maslow (1954, 1967). The more that elementary needs are satisfied, the more do questions such as that of personal growth become central. There could scarcely be a more thorough-going post-industrial value. Though scientific theories about personal growth are contemporary the concept itself is found in religions and philosophies which grew up in the pre-industrial world. It became marginal in the industrial world where the achievement emphasis was fostered by the Protestant Ethic. Now, sub-cultures focussing on the retrieval of self-actualization often use for this purpose Eastern religions and philosophies which have remained untransformed by industrialism.

Yet within the Western tradition medieval literature contains a supreme exponent of self-actualization in Dante who at the age of 35 went into the 'dark wood' to work through his mid-life crisis. The idea that personal development as distinct from personal achievement stops at adulthood is an industrial concept. One had to become adult to do the job but it was the job rather than oneself one then got on with. Achievement is a closed system, bio-physical rather than an open system, bio-social idea—as instanced in Freud's theory of genitality. But Erickson's work (1957) on identity has shown there are later identity crises than those of adolescence. This whole line of thought was foreshadowed by Jung's concept (1953) of individuation. Weathering the mid-life crisis was postulated as a critical process of self-renewal which determined capacity for forward movement in the 'third quarter of life'. Recent developments in psycho-analysis concerning the working through of conflicts associated with the paranoid and schizoid positions, regarded as continuous throughout life, have increased understanding of later as well as earlier possibilities for development (Klein, 1948; Jaques, 1965). The positive theory of madness advanced by Ronald Laing (1967) recovers the 'lost tradition' in a more recent socio-experiential idiom.

In the pre-industrial tradition capacity for personal growth beyond the attainment of psycho-sexual maturity was assumed to be a rare capacity reserved for the few—with inborn (or god-implanted) gifts in the arts, religion or philosophy. It was aristocratic. Maslow's researches convinced him it was widespread. The democratization of the value is the post-industrial phenomenon. Market researchers forecast increasing percentages of self-actualizing people in the American population in 1975, 1985—and, of course, in the year 2000.

Self-Expression

As regards self-expression, Friedman and Miller (op. cit.) report data obtained by the Stanford Research Institute showing that 50 million Americans now participate in some form of amateur act activity. 'Doing one's thing', a preoccupation which has become epidemic, represents a reaction against being forced to do someone else's thing—at work being coerced by narrowly prescribed roles, especially in a large organization; as a consumer being faced with the choice of only mass-produced goods; and as a social individual being expected to conform to one or other of a very limited set of life styles. Yet it is the affluence of the much criticized technological society that is producing a greater scope for choice. Self-expression is also concerned with trying out unused capacities, with discovering unsuspected potential. So far as the search for self-expression increases the variety and range of the individual's response repertoire it may be seen as aiding adaptation to more complex environments. But it cannot proceed far without raising questions about the identity of the self which is seeking 'expression'. Is this the true self, or 'am I my false self?'

Psycho-analysts have found that a great deal of the work in character analysis—and character analyses have become the most common type of psycho-analytic undertaking—consists in the undoing of a false self (Winnicott, 1958; Deutch, 1942). Denial of the deeper aspects and ordeals of self-actualization and self-expression will lead to the 'trivialization' of these values. Evidence of this is abundant in the cultism at present surrounding them.

Values of self-actualization and self-expression could lead to a solipsistic personalism were they not balanced by the other two values of the emerging set. Angyal (1958, 1966) regards the life process as consisting of two underlying, opposite, but complementary trends—which he calls autonomy and homonomy. Autonomy expresses the individual's need to separate himself from others, to establish his own domain and to expand this *vis-á-vis* his environment. Homonomy expresses his need to relate to others and to become part of something larger than himself, to do 'more than his thing', because after all he is not self-sufficient but belongs to a 'universe' which includes both himself and his environment. Self-actualization and self-expression embody the trend towards greater autonomy as do achievement and self-control. But the first two emphasize expansion of the person's being, rather than of his field of action, as do the second. Independence and endurance of distress continue the trend towards autonomy whereas inter-dependence and capacity for joy assert the countervailing trend to homonomy. The industrial values emphasize autonomy and the action frame of reference; the post-industrial homonomy and the existential frame of reference. The adaptive requirement of the future is a cultural system which embodies a balanced configuration of them all.

Inter-dependence

In developing a socio-experiential theory of human inter-dependence which recognizes the reality of the other as well as the self, Laing (1963) remarks that the word 'you' does not occur in Freud. So far as the ego has an id and a superego, all of which are contained within the person, he is right about the conceptual scheme which Freud developed. But he is wrong about the process of psychoanalysis which requires the presence of a real other in the person of the analyst. The development of child analysis by Melanie Klein (op. cit.) and Winnicott (1958, op. cit.) has increased the emphasis on an object-relations approach which recognizes the importance of the real object, or other, as well as the phantasied object, or other, and the interplay between them. For there can be no denying the importance of the real mother to the child in the situation of infantile dependence the psychological understanding of which has been extended by Bowlby (1969) from an ethological standpoint. Growing up, as Fairbairn (1952) puts it, is the conversion of dependence into inter-dependence. Capacity for inter-dependence will carry an ever higher premium in the post-industrial society, given that higher levels of complexity and uncertainty can only be met by the greater adaptive resources brought into being by self-regulatory collaborative endeavour.

Capacity for Joy

Inter-dependent life styles need to be founded on a high level of trust, which raises the question of the positive element in relations with, and feelings about, others. One cannot have good feelings for others unless one has a fund of good feelings in oneself. This means being on better terms with oneself than most people are—being less alienated from one's real self by a false self. When society may be construed as hostile and the work one must do for a living is replete with 'unpleasure'; when one's relations with others are competitive while the injunctions of the superego are forbidding: a philosophy of combative or stoical endurance is congruent with one's experience and needs. Tomkins (1962/63) points out that we are more experienced in regulating negative than positive affects. The culture of the work-addicted industrial society has been elaborated to aid adaptation to a negative affective life. When life is earnest we can deal with it; when it is not we are at a loss. The Grateful Dead (the quintessential San Francisco folk group) not only give their 1969 album a symbolically post-industrial title, 'The Working Man's Dead', but include a lyric the burden of which is 'having a hard time living the good life'—which would seem to be the core problem 'at the frontier'.

The prospect that the traditional distinction between work and play, made extreme in the industrial society, will be replaced by a new modality in which the reality and pleasure principles 'co-operate' is both incredible and terrifying. The infrastructure of distress, which has a numbing effect on the rest of the affective life, has functioned as a defence against the degree of self-encounter required to meet this challenge. What capacity does one have for joy when the world created by 'the dismal science' (the queen discipline of the industrial society) has issued an injunction against finding out?

Some of the contemporary sub-cultures assume that mere assertion of the capacity to love is enough to create at least a sub-world of joy–with badness and hate banished and displaced on to the 'brutal society' (even though Woodstock has had to reckon with Altamont and MacCarthy idealism with Weatherman terrorism). Gross projection carrying the assumption that we can enter our own heaven by getting rid of our own hell is indicative of the type of splitting which Emery and Angyal call superficiality. Moreover, hippie colonists and protesting extremists have been encouraged to be superficial by none other than the existential sages. Sartre himself has proposed that hell is other people. The new humanists (Maslow included) denounce the pessimistic world-view propagated by psycho-analysts supported, it would seem,by the establishment of the Judeo-Christian tradition (the establishment as regards the invention of 'sin'). Marcuse's doctrine of the conspiracy of repressive tolerance constitutes a social interpretation of Freud which is paranoid.

The fallacy is to suppose that love has a simple hedonistic basis–the pleasure principle. This leads to a sensate theory of the positive affects which can only end in solipsism. The experience of 'excitement' now becomes the summum bonum. Whether induced by another person, oneself or a drug is irrelevant; the sensation matters, the relationship does not; the other is an instrument to be used, not an object to be cherished. This value has obtained wide currency and is in favour with futurists (Kahn and Wiener, 1967). So far as it prevails joy cannot become linked to inter-dependence and allow homonomy to balance autonomy in a new value configuration.

If the version of psycho-analytic theory current before World War I supported a simple hedonistic theory of the positive affects, the version which began to develop in the inter-war period corrected this by the introduction of a new concept–*reparation* (Klein and Riviere, 1937). The first dilemma of the child is to love and hate the same needed but frustrating maternal object. His most primitive defence against the pain of his ambivalance is to split her (in phantasy and in his behaviour) into an idealized good fulfilling mother and a denigrated bad frustrating mother. In so doing he splits himself. Moreover in his greed for the good mother he exhausts her (to the point of losing her life in phantasy, her patience in reality); then turns her into a useless and therefore bad object who will revenge herself on him for what he has done to her. As the child's growing cognitive structure enables him to begin separating an external world from himself he begins also to realize that his good and bad mothers are parts of one real whole other person and his good and bad feelings parts of himself–another real whole person. The pain of this discovery is complicated by his having to admit that the goodness he so desperately needs and wants is hers not his (as he had assumed when innocent, i.e. in his original state of undifferentiated omnipotence). This arouses the deepest and most difficult to deal with of all the negative affects–envy. The child launches an envious attack on the goodness of the mother because the goodness is hers not his (to do what he likes with and to have always 'on tap'). Only through recognition of his predatory aggressions can the child find his sense of guilt and only through this his need to make reparation to the good mother he has begun to destroy (and will, in his anxieties, lose). But in making reparation he finds goodness in himself, in his good feelings for her, and in his gratitude for what she has given him. He cares for her and wishes to put her right. The elucidation of the envy-gratitude dynamics came very late in psycho-analysis (Klein, 1957). They show that the source of feelings of one's own goodness is in the experience of good feelings towards the other. This in turn involves recognition that the source of one's need satisfaction is in the other who has good feelings towards oneself. The experience of positive affects is founded on the recognition and experience of inter-dependence. These give the conditions for the person's realizing his independence; as he can only become whole (can only undo his own psychological splitting) by recognition of the contribution of the other.

Pleasure may be a component in good feelings but it does not bound their meaning. If good feelings are associated with sexuality, whether at the oral or the genital level, sexuality is an expression of inter-dependence and belongs to the homonomous rather than the autonomous trend. Goodness is discovered in the encounter with badness, not through its denial. Moreover, because good feelings are associated with making the other whole, they have an aesthetic component which is the basis of innovation (c.f. Ackoff, 1969a)—more than ever to be in demand in the post-industrial society. Plato excluded poets from the Republic because they were change agents. The need of a changing society is to strengthen these capabilities in as many people as possible.

But this means growing up, becoming more mature in the psycho-social as well as the bio-physical sense. Continued personal growth is indeed what is required but it can only come about by a working through of the deep ordeals which have always belonged to the human condition to a far greater extent and by far more people than in previous societies. We need more self-actualizing adults not more self-indulging children. All this means that the psycho-social price of survival in complex fast-changing environments is going up. We are not going to get away too much longer with having rather bad societies composed largely of rather immature people. To produce a better world turns out to be the hard job it is because it means becoming better oneself. To maintain the 'negative therapeutic reaction', not to change, is more comfortable. To remain in hell—where one lives a somewhat dreary life in a rather inhospitable atmosphere—is to remain on psychological easy street (however Satanic, or productive, the mill). To go further means a more thorough working through of the anxieties of the depressive position but it is this which releases the positive growth process. Such working through is unlikely to attain a new general level unless a new social context emerges; for the present level is a function of the present context. The regulation of positive affects requires the identification of new values and the formation of new norms. We need to find ways of going beyond the superego without falling back on the pleasure principle. In the making of a new society the psychologist's job is to go on showing that there is no salvation in the cultivation of superficiality as the basis of personal values. A society based on a denial of the deeper—and darker—aspects of the human psyche will be more one dimensional than the present and there will be no joy in it.

Chapter 14

Task and Contextual Environments for New Personal Values

Organizational Philosophies
Ecological Strategies
The Function of a Common Ground

Organizational Philosophies

New values will become salient only if experience in all parts of the life-space consistently supports their emergence. Therefore, what is happening in organizations (chapter 11) has particular importance in that it will affect the character of the task environments in which the individual is likely to find himself.

Column one of Table 2(b) summarizes classical organization theory and prevailing management practice, column two the emerging ethos.

Table 2(b) Organizational Philosophies

5. Mechanistic forms	Organic forms
6. Competitive relations	Collaborative relations
7. Separate objectives	Linked objectives
8. Own resources regarded as owned absolutely	Own resources regarded also as society's

It is in complex science-based enterprises, the most post-industrial (comprising the leading part, chapter 6), where the attitudes and values of column two are beginning to take hold.

Burns and Stalker (1961) showed the gathering influence of what they called the organismic pattern of management in the electronics industry as this sought to establish itself after World

War II in Scotland in an area in which were ingrained the autocratic and bureaucratic traditions of the first industrial revolution. The impact of McGregor's Theory Y (1960) in a number of countries on managers endeavouring to further the second industrial revolution illustrates the key importance of clarifying an emergent value. A new ideology arose in organization theory during the decade of the sixties which sought to identify a new organizational archetype—the trans-bureaucratic modality. Likert (1961) classified enterprises according to the number of steps taken in management style away from the traditional exploitative and authoritarian form. He has since been amassing evidence (1967) that his system of overlapping group hierachies (System 4)—the least exploitative and authoritarian—yields the best pay-offs in organizational performance. Argyris (1964) identified two steps 'further out' than Likert's System 4 by considering first the project form of organization which spread from R & D into production in the aerospace industry and next the need to renew sanction from a 'constituency' group when ground rules were changed. Bennis (1968) considered capability to belong, rapidly and effectively, to *temporary systems* not only as an organizational but as a life style congruent with a more mobile world. Social perception of the trans-bureaucratic mode as 'good' and of the bureaucratic mode as 'bad' would, as a posture, appear to be ahead of changes in practice. Likert thought it might take 10-15 years for a large organization to change its norms from System 1 or 2 to System 4. One is reminded of the danger that value-change may not take place at the pace required (chapter 4). 'Systems management' (chapter 6) has become very much the 'in mode'. Unfortunately, it still tends to be apprehended in a technical rather than a socio-technical frame of reference.

Competitive strategies appropriate to Type 3 environments predominate in what Galbraith (1967) has called the *New Industrial State* where, especially in the advanced industries, a few large firms, to varying degrees multi-national, vie with each other under conditions of oligopoly. Yet, as he points out, there are many indirect ways of calling off what would otherwise be a fight to the death without getting into the kind of trouble with the Anti-Trust Laws which ten years ago befell General Electric and Westinghouse. This incident was a warning, belated rather than early, that a Vickers type limit had been reached. For the student of emergent processes, it is not so much the growth of take-overs (a Type 3 phenomenon) which has been of interest during the last decade as the increasing tendency of large firms across the world to undertake 'joint ventures'. Joint ventures often combine public and private capital. Participants in a scheme recently put forward to develop a petro-chemical and shipyard complex in a part of Newfoundland suitable to take large tankers included the Provincial Government, the Federal Government of Canada, a British firm (B.P.) which is part public, part private, and various American private interests including Litton Industries. Such undertakings involve a strategy of 'shared parts' which is a Type 4 process. They involve substituting for an order based on the competitive challenge of superior power a *negotiated order* based on mutual accommodation of interests all regarded as legitimate. The term 'negotiated order' is surfacing in many fields at the present time: in the hospital world—concerning the relations of medical and other staff categories (Strauss, 1964); in the prison world—concerning staff-inmate relations (Rutherford, 1969); as well as in the labour-management world from which it came (Walton and McKersie, 1965). Vickers (1970) has contrasted 'negotiation' with 'containment' (by the use of coercive power) in the political evolution of industrial England. The idea of a negotiated order is congruent with the need to develop a greater capability to manage inter-dependence through co-operative rather than competitive relations (though not without confrontation).

The question of linked *vs.* separate objectives is also best approached by considering what has been happening to it in the leading part of the industrial field. In the values of the industrial society it was presumed that an enterprise had, and should have, an exclusively economic objective. The economic sector thrived best when its independence from other sectors was maximized. Though this idea was challenged many years ago in the management literature (Berle and Means, 1932) it received scant attention in practice until very recently. But the urban crisis in a country such as the United States is bringing home to corporations whose business inextricably binds their fate to the inner city that social losses can undermine economic gains. Externally, they have little option but to become actively concerned about the quality of the environment, just as internally they have to be actively concerned about the quality of work-life and career opportunity offered to employees at all levels. While most responses to all this are arrested at the level of tokenism, here and there more thorough-going changes are beginning to take place.

The issue is closely linked to that of attitude to ownership of resources which became a central topic in a project on company objectives and management philosophy which the Tavistock Institute began some five years ago with Shell Refining Company in London (1966). The company has since stated publicly that it regards its resources, both material and human, as belonging to society as well as to itself and has undertaken to manage these resources in accordance with this principle. The public statement was made only after the implications had been worked through over an 18-month period in a large number of residential conferences involving employees at all levels from the Board to the shop floor. Are resources to be regarded as owned absolutely (like slaves—or wives until not too long ago) or conditionally—and if the latter what are the conditions? Who are the claimants (other than oneself—or the stock holders)? What are their rights? How are they reconcilable?

This whole set of issues involves the concept of a negotiated order and the values—and competences—of inter-dependence. The following hypothesis seems warranted: *that so far as a negotiated order evolves it will reduce segmentation.*

Ecological Strategies

Signs of a new and adaptive trend are hard to discern in the contextual environment. This includes the wider world of large inter-organizational and inter-urban clusters. Yet the world of these clusters is sufficiently established in post-industrial societies for the systems of social ecology they compose to be regarded as their critical contextual element. The appropriateness of the strategies through which ecological as distinct from organizational systems are regulated becomes decisive in determining the adaptive effectiveness of the culture of post-industrialism.

In column one of Table 2(c) is set out the prevailing pattern, which reflects the persistence of an industrial concept of ecological regulation based on a combination of bureaucratic and laissez-faire principles. Column two contrasts it with a very different pattern which only in the last few years has begun to take shape. The list contains two sub-sets, items 9-13 being concerned with policy and items 14-17 with administration.

Table 2(c) Ecological Strategies

9. Responsive to crisis	Anticipative of crisis
10. Specific measures	Comprehensive measures
11. Requiring consent	Requiring participation
12. Damping conflict	Confronting conflict
13. Short planning horizon	Long planning horizon
14. Detailed central control	Generalized central control
15. Small local government units	Enlarged local government units
16. Standardized administration	Innovative administration
17. Separate services	Co-ordinated services

The clue to policy-making under industrialism is in the first item which discloses the passive mode of adaptation. A negative system of government was held to be the necessary counterpart of a positive system of enterprise. Intervention took place only when it became undeniable that a particular corrective was needed so that the specific measure became the unit of legislative action. Consent expressed in the vote of a simple majority of elected representatives (themselves chosen every four or five years by a similar method) constituted sufficient sanction, even if the majority was narrow. One did not mind being in the minority on some items and on some occasions so long as one could be in the majority on others. The pattern of an open series of specific measures restricted in scope, introduced piecemeal and sanctioned by a simple majority dampened conflict by allowing it to be absorbed in repeated small dosages. This pattern was inherently that of a 'non-planning' system. One kept accounts and saw that they balanced (deficit financing in any circumstances is still frowned upon in places such as British Columbia) but the budget year was the limit of the planning horizon. To look further ahead would have been to interfere with the future which was dangerous and immoral.

All this worked tolerably well so long as the main regulator at the ecological level—the market—was working auto-regulatively. The worst that could happen was a bad trade cycle swing. That of the Great Depression of the 1930s was so bad that it caused even the most *laissez-faire* societies to seek some form of economic regulation. Since then purely economic regulation has proved insufficient. Other dimensions of value have come to the fore. But these create incompatible demands which increase the dissociation of social sectors at a time when the need for them to become inter-dependent has increased (chapter 11).

So long as mixes of Type 1 and Type 2 conditions prevailed in the market place the relation between negative politics and positive economics could be maintained in balance. Increased salience of Type 3 conditions subjected this arrangement to increasing imbalance. Tactical accommodation has consisted of punctate intervention leading to jungle growth in the Administrative sub-set of Table 2(c). As Type 4 conditions become salient this model can be stretched no further. Reform at the administrative level in the direction of column two (repeatedly proposed) is impossible, beyond a modest limit, without re-orientation in the same direction at the policy level in terms such as those proposed by Lawrence Frank (repeatedly avoided) (chapter 7). It may be doubted whether the present culture of politics considered as a medium for the ecological regulation of society has the capability without modification to achieve this re-orientation. Should we start to behave consistently in terms of the policy items of column two we would have already changed our political character, even though we would not have re-written the constitution.

The familiar alternatives to democracies of the Western type are totalitarian regimes. These may be traditional in form, combining a coercive political with a stagnant economic system—as in pre-industrial societies—or they may be established as the result of revolutions, when they endeavour to combine coercive government with economic growth. Regimes of the first type have registered successes in delaying industrialization, regimes of the second in accelerating it. But these latter are societies begining industrialization. There is no evidence that totalitarian regimes can provide a safe passage to post-industrialism for advanced industrial societies. One may expect increasing disturbance in totalitarian countries the more complex they become.

The Function of a Common Ground

If political democracies of the Western type break down and if totalitarianism is even less able to meet complexity, in what direction may an answer lie? It would seem to lie in the direction of *social pluralism.* The question then becomes how much pluralism is containable, without further dissociation, unless a unifying element is also identifiable? One aspect of currently emerging pluralism is loss of the paramountcy of economic values. Other dimensions of value are re-asserting their claims. Even in the enterprise itself Lawrence and Lorsch (1967) have shown that better overall performance arises from an ability to accept a diversity of internal management climates appropriate to different functions. This is analogous to the acceptance of different life styles. Both involve increasing the range of one's directive correlations.

If we must learn to accept a variety of figures is there a common ground? We have not been accustomed to look at the 'ground'. Our training has made us figure-focussed. But a common figure we may not find. Common ground we may. The following hypothesis may be advanced: *that the capacity to accept the greater degree of pluralism that is characterizing the transition to post-industrialism, and which involves loss of paramountcy in any one value or 'figural' societal system, will be a function of the extent to which a unifying ground can become established.*

Marshall McLuhan (1964) can be of assistance in furthering our understanding of how this might be possible. He has shown how the new media consequent on the second industrial revolution are changing the socio-cultural 'ground forces' in a common direction, whatever the differences between individuals, groups and organizations. The conditions obtaining in Type 4 environments lead us to expect the emergence of more powerful ground forces than in simpler environments. These may produce unregulable disturbance as envisioned by Vickers. They may also produce a new type of background cohesiveness. In McLuhan's view they are altering the 'sense-ratios' from the dominance of visual to a re-emphasis of aural-tactile experience. They are altering simultaneously our cognitive structures from domination by an analytic to acceptance of a synthetic approach. We are learning to think and feel in terms of fields and simultaneity of processes as well as in terms of elements and causal trains. This is congruent with modern physics. Rational and non-rational modes of experience are not inherently incompatible.

Noticeable are the extent to which popular folk music is establishing itself as a lingua franca across a large part of the world and the unifying influence of the transnational youth culture. McLuhan adds that television being a 'cool' medium, increases involvement and decreases detachment.

Such processes have the effect of strengthening the trend towards homonomy. They do so by acting on the ground both in the environment and in ourselves rather than on the figures on which we focus. In this they are ecological. In time they may influence 'figural' appreciations in ways likely to be adaptive to post-industrialism. They are congruent with the forces impinging on the individual from other parts of his life space which have been tending to produce the emergent values which have been described. It may be hypothesized that *the establishment of more common ground will reduce dissociation.* Current experience, thence expectation, (outside a narrow range of social encounter) of lack of common ground inhibits exploration and increases isolation.

This potential for congruence may be noticed without our ceasing to notice also the disturbing effects of ground forces of another kind—those through which the limits of regulation are being reached. The appearance of a trend towards more background coherence constitutes a necessary but not a sufficient condition of successful adaptation to Type 4 environments. Active and constructive use will have to be made of it if one of the better of the alternative futures is to be attained. The communications technology of the second industrial revolution is available. As Emery has said (chapter 6) the consequential change which has taken place in intra-species communication is 'a greater mutation than if man had grown another head'. According to what we do with them, the new media may work for us rather than against us and increase the diffusion rate of new values not previously possible. At least they may prepare the ground.

Chapter 15

The Risk-Security Balance and the Burden of Choice[1]

The Personal Impact of Post-Industrial Trends
Income Trajectories—The Analysis of the Life Cycle
Emerging Changes in Personal Life Styles
The Impact of Developing Future Trends Upon Personal Financial Needs
The Burden of Choice

The Personal Impact of Post-Industrial Trends

That American society should be in a transition from industrialism to post-industrialism, from the economics of scarcity to the economics of abundance and subject to a much faster rate of change than any previously experienced is giving rise to different patterns of opportunities and constraints, of security and risk, than those to which we have been accustomed. It is profoundly changing the risk-security balance of the middle class individual. It will place a new burden of choice on many who will enjoy and suffer affluence for the first time in the terminal decades of the present century. How will they fare?

Affected are mundane matters such as financial arrangements. An examination of these will provide a perspective which can bring out more concretely than any other the extent to which the everyday life of the ordinary middle class person is exposed to the change processes discussed in this book. He will not escape them simply because he is not among the

[1] This chapter is based on the author's contribution to a pilot research project on the 'Future Environment of Financial Services', undertaken by the Management and Behavioral Science Center of the Wharton School of the University of Pennsylvania, and sponsored by the Insurance Company of North America. It was presented to the 1970 Research Conference of the American Society of Actuaries held at the Wharton School.

disadvantaged or the wealthy, the power elite, the dissenting vanguard or the withdrawing periphery. Of special interest in the diagnosis of his condition are his insurance needs, since insurance is the field specially concerned with risk and uncertainty. It may be suggested that the increasing salience of Type 4 conditions is changing the concept of insurance. On the one hand, it is being expanded to include much that it did not include before in the way of financial services, and on the other, being prompted to adopt a positive rather than a negative philosophy. By this is meant that insurance will have to become more concerned with the *maintenance of life opportunities,* both actual and potential, rather than simply with protection against loss and disaster. This is a consequence at the level of the individual of having to take the 'active role' (chapter 6).

The insurance requirements of the individual in personal lines are related to many factors. These relations can be represented by the following equation:

$$PL\ (In) = f(LS, LC, FS) \times I_{a+p}$$

where

In = Given Individual
PL = Requirements for Personal Lines of Insurance
LS = Structure and Content of the Life Space
LC = Phase of the Life Cycle
FS = Family Status and Type
I_{a+p} = Annual Income from All Actual and Potential Sources

The Changing Role and Nature of the Life Space. By life space is meant the world a person lives in as he perceives it at a given time. It is a combination of what is available to him in his environment and his construction of this in terms of his own needs, wishes and capabilities:

(i) This world has expanded. It is far larger than it used to be. There is far more in it. This will be even more so in the future.

(ii) It also contains many more kinds of things. It is becoming more variegated and differentiated, whether as regards material goods, personal resources, amenities, or the manifold array of opportunities and choices along any line of value (chapter 10).

(iii) These opportunities and choices are becoming more intricately interconnected. Many have elaborate prerequisites before they can become available, such as level of education.

(iv) The higher rate of change means that much is becoming more obsolescent more rapidly. There has to be a higher renewal rate and a capability to meet new and unexpected elements and conditions.

All in all, the world has become a far more stressful and demanding place and will become more so. To meet this the individual needs a more ample and more flexible repertoire of resources. His security consists in acquiring and maintaining this repertoire.

The Increasing Articulation of the Life Cycle. The life cycle used to be thought of in terms of infancy, childhood, adolescence, and then a single undifferentiated period of maturity till old age and death. Now the adult period is becoming differentiated into successive phases along several dimensions which may be present simultaneously:

(i) There is coming into being the serial career, with at least one major change around mid-point (of employer, often also of occupation or profession, or at least of the chosen line within an occupation) and, increasingly, another major change later.

(ii) There is the prolongation of education. Over 40% of even the present generation aged 18-21 have begun some form of college. Subsequently there are graduate studies for those seeking higher professional qualifications. This process is frequently not continuous and is mixed in awkward ways with the problem of getting launched on an initial career. Beyond this are the problems of continuing education, of keeping up and preparing oneself to seize new opportunities, e.g. at mid-career.

(iii) In parallel is the increasingly differentiated character of the family cycle: the early companionate, often temporary, marriage (during the student years); the establishment of the family of orientation with the coming of the first child; the increasing likelihood of divorce leading to the establishment of a second family with children; the larger number of post-child years.

All these aspects have considerable consequences for residential requirements, locational shifts and the type and amount of services a family needs.

Trends in the Nature of the Family. A person's family status is now to be thought of not simply as single or married; but as single, married, divorced, or remarried.

(i) Though the span of child-bearing years for women is contracting (with more and more women returning to employment), for remarried men the span of years of responsibility for children is increasing when second families arrive or are taken on.
(ii) Also, in addition to women working before and after the child-rearing years simply for economic reasons, there is a spreading pattern of the *double career family* (Rapoport, R. and Rapoport, R. N., 1969) where the woman works at a profession or analogous form of employment for interest and self-fulfilment. A considerable repertoire of services is required to keep such a family going.
(iii) Another problem is posed by the increasing number of three and four generation families, both with respect to economics and residence.

The more complex the form of the family, the greater the resources required to maintain it.

Trends in Total Personal Income. The patterns and changes described above, both present and to come, are not unexpectedly functions of the increasing affluence of the society (whether as cause or effect). The more complex life styles, with their changing and multi-dimensional sequences, cannot be attained or maintained except on the basis of a substantial and continuing personal and family income. An income of $10,000 a year (at 1965 dollars[2]) will be taken as the minimum level a family of average size must attain to be considered in any way

[2] All figures quoted throughout the chapter are in 1965 dollars. Amounts referring to the future are 'guesstimates' which represent the mean position taken in various recently published forecasts. An average family is a man, wife and two children.

affluent, and an income of $15,000 a year or more as a level at which the resource base becomes a little more ample. In 1965 some 25% of families had incomes of $10,000 or more; by 1975, 48% are expected to pass this threshold and by 1985, 65%. In 1965 only 7% had annual incomes of $15,000 or more, while by 1975 22% are expected to reach this level and by 1985 no less than 41%. Though the dates for these forecast percentages may be placed too hopefully near, a society is coming into existence in the foreseeable future in which the majority of the families will become affluent first by the $10,000 p.a. criterion, and later by the $15,000 p.a. criterion. Alongside the affluent sector is what may be called the sub-affluent sector–families with between $3,000 and $10,000 p.a.–the lower half of which comprises people who are exceedingly 'hard-up'. This comprised 59% of families in 1965 and is forecast as comprising 42% by 1975, and 31% by 1985. There is also a third sector composed of those in serious poverty, by the criterion that their annual income is less than $3,000. (Many would raise the 'poverty line' to $6,000). The third sector included 16% of the families in 1965, the estimated proportions for 1975 and 1985, respectively, being 10% and 4%. These figures take account, perhaps optimistically, of likely action in the public sector.

This brief analysis has shown the extent to which and the rate at which the affluent sector is likely to become dominant in the next two decades. By the year 2000 this affluence will have grown to a much greater degree. The previously dominant sub-affluent sector will, however, remain large for most of the present century. The third sector (the poor) will (it is to be hoped) become marginal in the United States as it already has in several of the smaller advanced countries, its need for protection being met by developments in the public sector.

The traditional ways of handling personal insurance and related financial services, in the form of a mass industry, have been fashioned to meet the needs of the hitherto dominant sub-affluent sector. The needs of the affluent sector are very different and can only be met by innovation. Though the methods used for the affluent sector when this was small may contribute, and some approaches may be carried over from the sub-affluent sector, neither source of experience is likely in itself to be sufficient (however necessary) to meet the needs of an affluent majority when this arrives, and whose affluence will increase. By the year 2000 the situation will be very different from what it was in 1970 and the level of affluence obtaining in the United States will to varying degrees be reflected in other advanced countries. (Japan may even be ahead, especially if a turn away from the values of the first industrial revolution gathers force in the U.S.).

Income Trajectories–The Analysis of the Life Cycle

The situation may be further examined with the assistance of Figure 1 which shows a series of income 'trajectories' through phases of the life cycle. The form of the curves is suggested by Jaques' work (1956, 1960) on earning progression curves. The incomes are total incomes and would include wife's and unearned income where these exist as well as the man's, supplements to which are likely to increase. The points of origin represent different levels of 'career launch' (one might say 'family launch') heavily influenced by level of formal education. The solid lines represent typical baseline income trajectories for each of four launch points. The sub-affluent trajectories are much lower than the affluent trajectories. With their steep ascent through mid-career, the affluent trajectories are vulnerable to a wide range of vicissitudes (indicated by dotted lines in Figure 1). A change by ± $2,000-$3,000 p.a. may make a critical difference to the quality of life at the sub-affluent levels. This must also be considered.

A number of perturbations of the baseline trajectories may be encountered. Some examples are illustrated by the dotted paths in Figure 1. The following characteristic features may be noted:

Reinforcements. These are additions to the breadwinner's salary from a wife's earnings (often substantial and increasingly equal) or funds available through one mechanism or another including securities, estate and parental sources. Their role is to raise the trajectory from career launch to mid-career. This has the effect of (1) making the life cycle less vulnerable and (2) making possible a line of development on a higher trajectory. Only committed and indefatigable entrepreneurs will attempt unaided the steepest trajectories. One may expect fortune-making to remain an obsession for some, if not for many. The

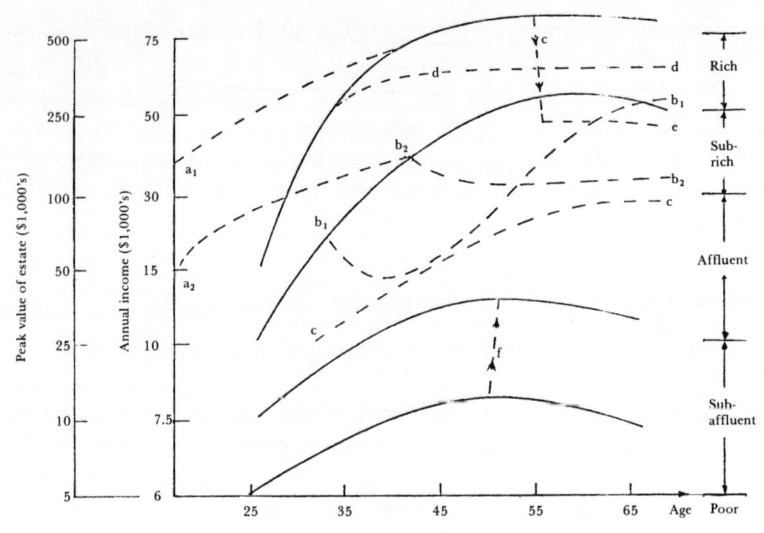

Figure 1 *Income Trajectories on the Family Life Cycle.*

KEY

a_1, a_2 Reinforcements at Different Levels.
b_1, b_2 Set-Backs With and Without Recovery.
c Delays.
d Flattening.
e Reductions.
f Switches.

double career family is likely to become far more important as a means of trajectory raising.

Set-backs. These are events which pull the individual and his family off the trajectory which they have come to assume as representing the socio-economic line of their life cycle. All plans tend to be made in terms of this line, which is more implicit than explicit. Nevertheless, expectations are guided by deep intuitions and powerful norms. Set-backs, which of course can be repeated, are of two kinds: those from which there is a recovery and those from which there is no recovery (recovery, of course, being often a matter of degree). The following properties of these events are relevant:

(i) The earlier the set-back, if serious, the more far-reaching the likely consequences.

(ii) The higher the trajectory, the deeper the possible descent.

(iii) The higher the trajectory, the longer the recovery time and the greater the likelihood of recovery being only partial.

(iv) Mechanisms exist which can arrest set-backs such as reinforcements and other forms of protection which can be acquired. These are likely to gain in importance as multi-valued choices increasingly affect life styles.

Delays. In view of the steepness of ascent of the higher trajectories (without reinforcements), delays—especially delays in first career launch—can have seriously negative effects. Such delays are intimately connected with the rate at which graduate education is completed. If a person is not launched on a first main career until he is well into his thirties as distinct from being still in his late twenties, he will probably proceed on a lower trajectory than he otherwise would.

Flattening. A not uncommon occurrence is for people who have started well to fade out. This is often associated with the way the mid-life crisis is handled. Jung (1953, op. cit.) for example, found that a lot of patients around the age of forty would come to him and simply say they were 'stuck.' Processes of self-renewal are profoundly necessary if a person is to go 'onward and upward' into the 'third quarter of life.' What he must do if he is to continue his personal development (self-actualization) is to maintain his capability for innovation. This may entail a 'psychological moratorium,' going into retreat for a while (often taking the form of going back to 'school'), taking an entirely new job, facing a divorce, etc. He must take risks in order to move forward. An individual is more likely to be able to take these sorts of risks if he has a reasonable degree of security.

Reductions. These are different in quality and implication from flattening and may, for example, be a function of a positive resolution of the mid-life crisis. The outcome may be that the individual chooses a life style in which he no longer seeks to increase his income. He may accept a lower level of income in order to secure other benefits. One may expect an increase in such decisions in the decades ahead. Such decision-makers will need to make provisions in one form or another if they are to do this without damage to members of their families.

Switches. These are the opposite of reductions—the individual succeeds in switching from a lower to a higher trajectory in a step-function way. This process may be a matter of luck or of seizing newly-presented opportunities which were not there at an earlier period (e.g. through educational delay or a wife contributing a second income); or it may be an outcome of the way in which later identity crises are resolved. The contributing factors are numerous, but one may postulate an underlying change in the risk-security balance of the individual. One would expect switches to be more important in crossing the boundary from sub-affluence to affluence than from affluence to riches, as the economic gains would make a wider range of non-economic choices available for the first time. The possibility of switching could become an objective in certain forms of sub-affluent planning, as a central norm of the majority society is likely to be concerned with exercising choice in the matter of life style.

Emerging Changes in Personal Life Styles

Certain features of the life styles associated with these emerging patterns will tend to be organized differently from the way they have been entered into in more familiar life styles. These differences are likely to be accentuated as time goes on.

From Succession to Simultaneity. Formerly a person went to school; then worked; then retired. Learning, working, and leisure were successive. Now, more and more people are living a life of learning, working and leisure at the same time. The leisure is to be thought of as required recreational opportunity. It is necessary to have 're-creative' experiences if one is to work 'creatively.' The work that men, as opposed to machines, are required to do (as the automation/computer technology spreads) will be a creative activity rather than alienating drudgery. The capability to do this kind of work entails continuous learning as well as recreation. Therefore, the individual must maintain his opportunities in learning and leisure. One way of phrasing this is to say that 'jobs' are giving place to *life roles* and that what one wants to 'insure' is a way of life.

From Stasis to Mobility. More and more people are not staying put. More and more of a person's affiliations are to systems which are in varying degrees temporary. Bennis and

Slater's recent book entitled *The Temporary Society* (1968) analyses this trend. Less and less are jobs, careers, and homes expected to last for life. Formerly, the advantages seemed to follow from getting one's self built in; now the disadvantages seem to grow from being locked in. A major need in the on-coming decades is to maintain one's capacity for mobility. But keeping options open is expensive. Mobility must be planned for and provided for. How can one insure one's mobility?

From Owning to Leasing and Renting. Simultaneous learning, working and recreation, together with maintained mobility, require a substantial extension of the environment in which a man and his family conduct their lives. The various parts of this environment are becoming less linked in space and time. Moreover, a good many of the contents change from time to time—whether jobs, schools, houses, cars, boats, vacations, cultural pursuits or various types of special equipment. There is a limit to the practicality of managing such an environment by directly and permanently owning as many as possible of the main items required by several different sets of conditions existing simultaneously and successively in the phases and situations of the life cycle. What a man needs to 'own' is the *structure* of the life-space he requires, not the items in it. Yet owning the items has been the traditional pattern. Recent trends suggest a strategy of leasing or renting the items in order to maintain control over the structure. This trend is likely to grow.

From Goods to Services. What people need in these evolving life styles is not so much more goods as more services. To give one example, a double career family with young children is in special need of household help which cannot be provided by the obsolete method of having servants. Other examples could be given from any field of family concern. What people want to do is to spend their time on their own priorities in work, learning, and leisure. They do not want to have their time wasted by having to attend to other matters which they are willing to pay to have done; and they object when services offered are unreliable and still more time is wasted. One may expect a rising trend away from the purchase of goods to the purchase of services. This is one method by which greater control over the structure of the life space may be obtained.

The Impact of Developing Future Trends Upon Personal Financial Needs

The implications for the financial management of the individual's life cycle in relation to his family enterprise must now be considered. The family, no less than the business enterprise, has alternative futures. Whether or not one of the more desirable of the possible futures is attained at any stage in the life cycle depends in the family, as much as in the business case, on the relevant information being available and on the quality of the planning. At least this is so as soon as the assets to be deployed become appreciable. Modest as the scope may be in the sub-affluent categories, it rapidly widens as the threshold of affluence is crossed.

The Nature of the New Game. As in business, everything goes wrong if a man thinks he is in one kind of business when he is really in another. He then fails to recognize the character of the changes which have been taking place. The family with assets to deploy, especially in a fast changing, uncertain environment, is no longer in the minimum game of protecting itself against certain commonly recognized forms of loss (including the death of the breadwinner). It is in the more ambitious and risky game of maximizing opportunities, given the resources available, to engage in the new life styles. It is the capability to maintain these opportunities at a level which allows a margin of safety that has to be 'insured'. Once this capability is seriously impaired, the chances of full recovery from set-backs are small: trajectories become flatter; the options go; mobility is diminished. Security in the new sense is lost.

The Rules of the New Game. What a person now has to do is simultaneously to generate, maintain, and consume his estate. It is as if he had already to be living after himself in order to live now. To be able to do this he requires *working capital.* A person needs to deploy and maintain this working capital in relation to the changing requirements of his family and himself. He particularly needs it in the earlier phases of his life cycle up to and through mid-career, when he needs funds which he can get hold of on short notice in order to undertake emergent missions.

The Integration of Financial Services. It is unlikely that the flexibility required can be obtained unless the different forms of financial service available to the individual and his family can become more integrated. This suggests the desirability of comprehensive but flexible arrangements which permit linkage and transfer without subjecting the individual to a system which establishes tight monopolistic control over his life. A systematic examination of existing ways in which this is being attempted and a search for improved ways in which it might be attempted require a sustained research effort.

Counselling. The bulk of the families who will be deploying substantial assets in the next decades will have little experience with sophisticated financial management of their private affairs. To do this well is always difficult; one is too closely involved emotionally with the issues and the people—who are one's own. But with a background of inexperience, some form of professional counselling from the outside becomes indeed an imperative. The design of a set of financial services which will offer the requisite flexibility security combination and of a set of professional roles to mediate them represents a long-term undertaking requiring the participation of many disciplines inside and outside the financial service industries. Emergent professionalism in this area will need to dissociate itself from selling; accreditation and fees will both be necessary.

The Burden of Choice

Individuals and families who cross the boundary of affluence, and these will become the majority in the decades ahead in a country such as the United States, must learn to live with the burden of choice. This is the price to be paid for emancipation from the constraints of hardship.

A number of informal group discussions and interviews (with 50 such respondents) conducted as a pilot undertaking for the project which gave rise to this chapter revealed a degree of bewilderment (across generations and levels of education) which was producing a paralysis of decision-making in a quarter of the respondents and considerable worry in another third. Among those who felt they were coping several recognized the problem.

Many of the respondents were uncomfortable because they sensed there were more options open to them than they could see. They did not know where to get relevant information. The implications of different courses of action became too complex to work out. Uncertainty regarding future developments sometimes removed any basis for calculation.

Apart from the difficulties experienced in making reality judgements several respondents said they had come to realize that they did not, as they supposed, have a set of values to guide them in making the choices they now had to make. This was because these choices involved what Ackoff (1969b) has called stylistic as distinct from performance objectives. They were not sure that they would like living with the consequences of their decisions—they had no experience of what was involved—yet they had to make commitments which could not easily be withdrawn.

Though the talk in these discussions and interviews was about personal affairs the texture as regards concern with uncertainty and complexity was similar to that which in an organizational setting formed a background of concern in the Shell management philosophy project (chapter 14). It also reminded the writer of much that was said at a recent conference on government planning which he attended in Britain. The beginning of the sixties decade was recalled as a time when there was a will to plan but little technical capacity. At the end, though technical capacity had increased, the will had gone. A number of matters had become so complex, and there was so much uncertainty, that one might as well leave them to themselves. One did not know how to intervene.

These are the difficulties of adapting to Type 4 environments. They pervade all system levels and all domains of concern. They express the bewilderment of all those accustomed to a simpler and more certain world, who now face the task of learning a new art of living, and who are in search of new guide lines, both intellectual and emotional, and on the personal, organizational and broader societal planes. How far and how fast to move towards the modalities in column two of Tables 2(a), (b) and (c) in chapters 13 and 14?

From Planning Towards the Surrender of Power

Planning as Process
Pluralism
A Required Dissociation
Re-instating Pre-history in Post-history

Planning as Process

The structural-cultural mis-match which has been examined, together with the nature and direction of the emergent adaptive processes, has led to a re-statement of Emery's position—that we must take an active, intervening rather than a passive, respondent role in regulating the welfare and development of society. This means facing the planner's dilemma. How can planning be carried out under conditions of accelerating change and rising uncertainty? What style of planning, if any, provides a field of manoeuvre for getting out of ecological traps?

For most people planning conjures up the idea of comprehensive planning, technocratically devised and centrally imposed. Such planning rests on two assumptions:

(a) that there was once a steady state and that there will again be another steady state—and that the way to get from the first to the second is to produce a complete plan which supposes that the principal future states of the system can be foreseen;

(b) that implementation can be carried out with resources completely under one's own control.

This represents closed system thinking, the machine theory of organization, the maximization of power—everything which the encounter with higher levels of complexity and uncertainty has shown to be unworkable. Yet a relapse into non-planning is equally unworkable. A form of anticipation and regulation has to be found which will match the realities of rapidly changing Type 4 environments.

Fortunately an approach to planning which has some of the requisite characteristics has begun to make an appearance in theory and in practice. It is an open-ended type of planning which takes account of the emergent, allows for the unforeseeable and follows strategies of inter-dependence rather than independence in the mobilization of resources. It has acquired various names: innovative planning (Friedman and Miller, 1965); interest group planning (Chevalier, 1968); adaptive planning (Ackoff, 1969b). The 'disjointed incrementalism' proposed by Hirschman and Lidblom (1962) would be regarded by many as a first conceptual move in this direction (chapter 6) yet in some contexts it could be a 'reaction formation' against comprehensive planning. It would then be indistinguishable from 'muddling through', which worked only when auto-regulative processes were functioning at the ecological level and the specific measure could be the unit of legislative action. The type of planning now required must be able to give direction and to set standards in the sense asked for by Frank (chapter 9) while at the same time being flexible and always remaining incomplete.

Thanks to Monnet, some French planners began fifteen years ago to put this type of planning into practice based on recognition of a simple but fundamental truth: that planning is not so much a programme as a process. However technical many of its aspects may be, in underlying nature planning is a social process. Moreover it is continuous. Phases of formulation, implementation, evaluation and modification succeed and interact with each other without reaching a final limit. It is also participative. All those concerned must contribute in appropriate roles. Else, one may ask, regarding continuity, 'what of the plan, now that circumstances have changed?' and, regarding participation, 'who is making plans about what for whom?'

Michel Crozier (1966), in carrying out a sociological study of the decision-making process of the Ve Plan, has shown that the most important effect of French planning lies not so much in the achievement of the targets as in the social learning released in the innumerable commissions which take part in making, carrying out, and revising 'the plan'. Each of these commissions represents a particular coalition of interests, often never brought together before, or previously too antagonistic to co-operate. This whole process, whatever its shortcomings (it has been too élitist even though cross-élitist), got going during the Fourth Republic when the party political system was collapsing and the Algerian war dividing the country. It has continued during the Fifth Republic. De Gaulle had to take the need for more involvement into account in the 1968 crisis of his country. But the nature of the emergent social processes remained unrecognized. A regressive characterization was made in terms of old 'images' which presented the choice as that of retaining the familiar 'bourgeois' society or facing 'communism'—rather than taking innovative steps in the transition towards post-industrialism. A moment of opportunity was lost (though promises were made) to begin educational reform in the thorough-going way the society required. No attempt was made to find a possible new meaning in the unusual and precarious coalition which emerged between students and workers. This was discussed in an interview given by Sartre to Cohn-Bendit. Sartre took the position that because being a student represented a temporary status while being a worker represented a permanent status there could be no coalition. Cohn-Bendit pointed out that in the on-coming society being a student would indeed become a permanent status because learning had become life-long, while all workers would need the opportunity to go on learning. Though Sartre took the point, the implications were lost in an ideological jumble which mixed Trotskyism, Maoism and Dadaism and in which the past is easier to discern than the future.

Nevertheless, so far as planning as a continuous and participative activity becomes organically linked to the open-ended political process of a society, this would seem to become more 'robust'—more able to survive the knocks and shocks of rapidly changing environments. The hypothesis is advanced that *the planning process (not the plan) can become the basis of a new 'culture' of politics.* This new political culture will involve continuous dialogue, painful confrontation, 'animation sociale', hard bargaining and multiple interest group accommodation. But these are the processes which can lead to innovative joint problem-solving as experience is gained and greater trust is established. The new culture will also demand a full technical 'input' from the planning professions and their supporting sciences at all stages and levels. There is likely always to be a certain dissonance between the technical and the participative aspects, and the 'planner's dilemma' is never likely completely to be resolved, though attendant conflicts, we may hope, will on the whole be benign rather than malign (in Boulding's sense) (1966).

Pluralism

The task of the new politics will be the regulation of ecological systems undergoing rapid but uneven change. One of the troubles is that there is a galaxy of such systems which overlap in various dimensions. Somehow we must find a way of reducing both systems and dimensions to a manageable set—though neither by opportunistic circumscription nor over-centralized control. A greater coherence in the 'ground forces' will strengthen capability to tolerate a greater degree of social pluralism in 'figural' systems. *It is hypothesized that the development of social pluralism may represent a method of complexity reduction without regressive simplification.*

To succeed in a problem-continuing environment post-industrial politics must become both more informed and more participative than the politics of industrialism, more devolved and open to more rapid and continuous feedback. Post-industrial man will spend more of his time in politics than industrial man and more in the planning processes associated with it. He may be presumed to have the leisure.

An anti-planning ideology persisting from the most powerful industrialism in the world has delayed the emergence of acceptable forms of the planning process in North America—the herculean and often competing efforts of special agencies on the more discrete 'industrial' model notwithstanding. Now that they have begun to appear the context is not so much as in Western Europe that of the national economy as that of the urban environment, where the central meta-problem lies. Relationships between planning and politics which actively involve the citizenry in all this, and which do not exist at present, need to be evolved. They will have to be pluralistic, at all levels—local, state, provincial, and national—and between government and non-government organizations. Though few can be sanguine, a thread of progress would seem to be discernible despite serious vicissitudes, regressions and tragedies of assassination in the United States and more recently in Canada. Several new types of coalition between public and private power and statutory and voluntary bodies have begun to appear, if uncertainly. Though crash programmes have failed, a number of small projects have produced a first growth of involvement. Some experience has been gained in the harnessing of protest, despite mounting violence. Some professionals and academics, however slowly, are beginning to work out new roles which are enabling them to retain their commitment to professional and scientific values while collaborating with the other actors.

A Required Dissociation

Nevertheless, the United States—with super-power responsibilities, a population of 200 million and an heterogeneity which has made extreme the processes of internal fragmentation, segmentation and dissociation which to varying degrees have affected all advanced countries—is a society too complex to tolerate much innovation in the social present in the political engagement of the planning process. Canada is better placed, being a very advanced country yet out of the storm centre of world politics; and possessing a highly educated but relatively small population. These conditions make more manageable some of the problems critical for the transition to post-industrialism. However severely Canada may be challenged by problems such as that of Quebec, these problems do not seem as intractable as many in other parts of the world. In recent decades the Scandinavian countries have been responsible for several social and cultural innovations in the European context. All may not be affluent in these countries but poverty has been eliminated. The stability and coherence of the ground has permitted such figural innovations as industrial democracy in Norway, the containment of pornography and the mega-family in Denmark (in which the Church played a leading part), prison reform in Sweden and an aesthetic renaissance in architecture and design which has spread through all these countries. Canada has some of the properties of a North American Scandinavia. U.S. innovation is in particular places.

Earlier in this book (chapter 9) the suggestion was made that some of the smaller advanced countries may become social laboratories from which others may learn in the transition to post-industrialism. One reason for this is that they are experienced in not exercising political power on the world stage, while remaining economically competent in the world market. They proclaim the viability of a dissociation between two elements on whose association the major states not only of the industrial but the pre-industrial period have been founded. To amass power has been the strategy of choice in adapting to a world in which Type 3 environmental conditions have been salient. *To surrender power is a necessary condition for survival in a Type 4 environment.* The Great Powers and the Super-Powers have found (and find) themselves obliged to spend unduly large proportions of their resources and energies in elaborating and maintaining specialized control systems (military and other)—in accordance with Emery's first design principle. Sooner or later (as the history books never tire of recounting) this weakens their core societies. By contrast the smaller advanced countries are closer to Emery's second design principle of self-regulation. They have preserved a core identity while maintaining themselves trans-nationally. They have successfully cultivated well selected capabilities in the economic field which embody cultural traditions which have grown out of their whole ecological setting and in which they have comparative advantage.

Re-instating Pre-history in Post-history

In certain respects the world of the post-industrial society, if this manages to establish itself, will recover some of the properties of the Type 2 environment. Anthropologists, perhaps, see this more clearly than others. Levi-Strauss (1960) has spelled out the argument in his Inaugural Lecture on being elected to the first Chair of Social Anthropology at the Collége de France:

> So-called primitive societies . . . have specialized in ways different from those which we have chosen. Perhaps they have, in certain respects, remained closer to very ancient conditions of life, but this does not preclude the possibility that in other respects they are farther from those conditions than we are.
>
> . . . Those which have best protected their distinctive character appear to be societies predominantly concerned with persevering in their existence.

The way in which they exploit the environment guarantees both a modest standard of living and the conservation of natural resources. Their marriage rules, though varied, reveal to the eye of the demographer a common function, namely to set the fertility rate very low and to keep it constant. Finally, a political life based on consent, and admitting of no decisions other than those unanimously arrived at, seems conceived to preclude the possibility of calling on that driving force of collective life which takes advantage of the contrast between power and opposition, majority and minority, exploiter and exploited.

In a word, these societies, which we might define as 'cold' in that their internal environment neighbours on the zero of historical temperature, are, by their limited total manpower and their mechanical mode of functioning, distinguished from the 'hot' societies which appeared in different parts of the world following the Neolithic revolution. In these, differentiations between castes and between classes are urged increasingly in order to extract social change and energy from them.

When, on the morrow of the Neolithic revolution, the great city-states of the Mediterranean Basin and of the Far East perpetrated slavery, they constructed a type of society in which the differential statuses of men—some dominant, others dominated—could be used to produce culture at a rate until then inconceivable and unthought of. By the same logic, the industrial revolution of the nineteenth century represents less an evolution oriented in the same direction, than a rough sketch of a different solution: though for a long time it remained based on the same abuses and injustices, yet it made possible the transfer to *culture* of that dynamic function which the protohistoric revolution had assigned to *society*.

If—Heaven forbid!—it were expected of the anthropologist that he predict the future of humanity, he would undoubtedly not conceive of it as a continuation or a projection of present types, but rather on the model of an integration, progressively unifying the appropriate characteristics of the 'cold' societies and the 'hot' ones. His thought would renew connections with the old Cartesian dream of putting machines, like automatons, at the service of man. It would follow this lead through the social philosophy of the eighteenth century and up to Saint-Simon. The latter, in announcing the passage 'from government of men to the administration of things', anticipated in the same breath the anthropological distinction between culture and society. He thus looked forward to an event of which advances in information theory and electronics give us at least a glimpse: the conversion of a type of civilization which inaugurated historical development at the price of the transformation of men into machines into an ideal civilization which would succeed in turning machines into men. Then, culture having entirely taken over the burden of manufacturing progress, society would be freed from the milliennial curse which has compelled it to enslave men in order that there be progress. Henceforth, history would make itself by itself.

This statement, which at one and the same time presents an analysis and a vision, conceives the viable future as an achievement of societal self-regulation made possible by the culture of the second industrial revolution. This would bring to an end the epoch we have known as history and have called civilization. Yet there can be no simple return to a Type 2 world in which large numbers of small societies existed independently; rather will there come into being a matrix in the inter-dependent networks of which values will be such that pluralism can be tolerated. But, if diversity is to be internalized and to become figural, similarity will be needed as external support in a common ground. The coolness which McLuhan discerns in the new cultural media may enable coldness as conceived by Levi-Strauss to re-enter society. Somehow the heat has to be taken off and the self-exciting systems described by Vickers brought under regulation. If planning processes are to be a principal means by which this is brought about, planning would, as Ackoff has suggested, work itself out of a job—if ever an ideal state of self-regulation could be reached in the world as a whole.

The Socio-technical System as a Source Concept[1]

From Closed to Open Systems
The Technological Component
Boundary Keeping
The Environment

From Closed to Open Systems

The analysis of the characteristics of enterprises as systems would appear to have strategic significance for furthering our understanding of a great number of specific industrial problems. The more we know about these systems the more we are able to identify what is relevant to a particular problem and to detect others that tend to be missed by the conventional framework of problem analysis.

The value of studying enterprises as systems has been demonstrated in the empirical studies of Blau (1955), Gouldner (1954), Jaques (1951), Selznick (1957) and Lloyd Warner (1947). Many of these studies have been informed by a broadly conceived concept of bureaucracy, derived from Weber and influenced by Parsons and Merton:

> They have found their main business to be in the analysis of a specific bureaucracy as a complex social system, concerned less with the individual differences of the actors than with the situationally shaped roles they perform. (Gouldner, op. cit.)

Granted the importance of system analysis there remains the important question of whether an enterprise should be construed as a 'closed' or an 'open system', i.e. relatively 'closed' or 'open' with respect to its external environment. Von

[1] This Appendix is a shortened version, empirical detail omitted, of a joint paper presented at the 6th International Conference of the Institute of Management Sciences, Paris, 1959. The original appeared in *Management Sciences: Models and Techniques*, Churchman, C. West and Verhulst, M. (eds.) (1960), London: Pergamon Press.

Bertalanffy (1950) first introduced this general distinction in contrasting biological and physical phenomena. In the realm of social theory, however, there has been something of a tendency to continue thinking in terms of a 'closed' system, that is, to regard the enterprise as sufficiently independent to allow most of its problems to be analysed with reference to its internal structure and without reference to its external environment. Early exceptions were Rice and Trist (1952) in the field of labour turnover and Herbst (1954) in the analysis of social flow systems. As a first step, closed system thinking has been fruitful, in psychology and industrial sociology, in directing attention to the existence of structural similarities, relational determination and subordination of part to whole. However, it has tended to be misleading on problems of growth and the conditions for maintaining a 'steady state'. The formal physical models of 'closed systems' postulate that, as in the second law of thermo-dynamics, the inherent tendency of such systems is to grow toward maximum homogeneity of the parts and that a steady state can only be achieved by the cessation of all activity. In practice the system theorists in social science (and these include such key anthropologists as Radcliffe-Brown) refused to recognize these implications but instead, by the same token, did '*tend* to focus on the statics of social structure and to neglect the study of structural change' (Merton, 1949). In an attempt to overcome this bias, Merton suggested that 'the concept of dysfunction, which implies the concept of strain, stress and tension on the structural level, provides an analytical approach to the study of dynamics and change'. This concept has been widely accepted by system theorists but while it draws attention to sources of imbalance within an organization it does not conceptually reflect the mutual permeation of an organization and its environment that is the cause of such imbalance. It still retains the limiting perspectives of 'closed system' theorizing. In the administrative field the same limitations may be seen in the otherwise invaluable contributions of Barnard (1948) and related writers.

The alternative conception of 'open systems' carries the logical implications that such systems may spontaneously re-organize toward states of greater heterogeneity and complexity and that they achieve a 'steady state' at a level where they can still do work. Enterprises appear to possess at least these characteristics of 'open systems'. They grow by processes of internal elaboration and manage to achieve a steady state while doing work, i.e. achieve a quasi-stationary equilibrium in which the enterprise as a whole remains constant, with a continuous 'throughput', despite a considerable range of external changes (Lewin, 1951).

The appropriateness of the concept of 'open system' can be settled, however, only by examining in some detail what is involved in an enterprise achieving a steady state. The continued existence of any enterprise presupposes some regular commerce in products or services with other enterprises, institutions and persons in its external social environment. If it is going to be useful to speak of steady states in an enterprise, they must be states in which this commerce is going on. The conditions for regularizing this commerce lie both within and without the enterprise. On the one hand, this presupposes that an enterprise has at its immediate disposal the necessary material supports for its activities—a workplace, materials, tools and machines—and a work force able and willing to make the necessary modifications in the material 'throughput' or provide the requisite services. It must also be able, efficiently, to utilize its material supports and to organize the actions of its human agents in a rational and predictable manner. On the other hand, the regularity of commerce with the environment may be influenced by a broad range of independent external changes affecting markets for products and inputs of labour, materials and technology. If we examine the factors influencing the ability of an enterprise to maintain a steady state in the face of these broader environmental influences we find that:

(a) the variation in the output markets that can be tolerated without structural change is a function of the flexibility of the technical productive apparatus—its ability to vary its rate, its end product or the mixture of its products. Variation in the output markets may itself be considerably reduced by the display of distinctive competence. Thus the output markets will be more attached to a given enterprise if it has, relative to other producers, a distinctive competence—a distinctive ability to deliver the right product to the right place at the right time;

(b) the tolerable variation in the 'input' markets is likewise dependent upon the technological component. Thus some enterprises are enabled by their particular technical organization to tolerate considerable variation in the type and amount of labour they can recruit. Others can tolerate little.

The two significant features of this state of affairs are:

(i) that there is no simple one-to-one relation between variations in inputs and outputs. Depending upon the technological system, different combinations of inputs may be handled to yield similar outputs and different 'product mixes' may be produced from similar inputs. As far as possible an enterprise will tend to do these things rather than make structural changes in its organization. It is one of the additional characteristics of 'open systems' that while they are in constant commerce with the environment they are selective and, within limits, self-regulating;

(ii) that the technological component, in converting inputs into outputs, plays a major role in determining the self-regulating properties of an enterprise. It functions as one of the major boundary conditions of the social system of the enterprise in thus mediating between the ends of an enterprise and the external environment. Because of this the materials, machines and territory that go to making up the technological component are usually defined, in any modern society, as 'belonging' to an enterprise and excluded from similar control by other enterprises. They represent, as it were, an 'internalized environment'.

Thus the mediating boundary conditions must be represented amongst 'the open system constants' (von Bertalanffy, op. cit.) that define the conditions under which a steady state can be achieved. The technological component has been found to play a key mediating role and hence it follows that the open system concept must be referred to the socio-technical system, not simply to the social system of an enterprise.

The Technological Component

It might be justifiable to exclude the technological component from the system concept if it were true, as many writers imply, that it plays only a passive and intermittent role. However, it cannot be dismissed as simply a set of limits that exert an influence at the initial stage of building an enterprise and only at such subsequent times as these limits are overstepped. There is, on the contrary, an almost constant accommodation of stresses arising from changes in the external environment; the technological component not only sets limits upon what can be done, but also in the process of accommodation creates demands that must be reflected in the internal organization and ends of an enterprise.

Study of a productive system therefore requires detailed attention to both the technological and the social components. It is not possible to understand these systems in terms of some arbitrarily selected single aspect of the technology such as the repetitive nature of the work, the coerciveness of the assembly conveyor or the piecemeal nature of the task. However, this is what is usually attempted by students of the enterprise. In fact:

> It has been fashionable of late, particularly in the 'human relations' school, to assume that the actual job, its technology, and its mechanical and physical requirements are relatively unimportant compared to the social and psychological situation of men at work (Drucker, 1952).

Even when there has been a detailed study of the technology this has not been systematically related to the social system but been treated as background information (Warner, op. cit.).

In the earliest Tavistock study of production systems in coal mining it became apparent that 'So close is the relationship between the various aspects that the social and the psychological can be understood only in terms of the detailed engineering facts and of the way the technological system as a whole behaves in the environment of the underground situation' (Trist and Bamforth, 1951).

An analysis of a technological system in these terms can produce a systematic picture of the tasks and task interrelations required by a technological system. However, between these requirements and the social system there is not a strictly determined one-to-one relation but what is logically referred to as a correlative relation.

In a very simple operation such as manually moving and stacking railway sleepers ('ties') there may well be only a single suitable work relationship structure, namely, a co-operating pair with each man taking an end of the sleeper and lifting, supporting, walking and throwing in close co-ordination with the other man. The ordinary production process is much more complex and there it is unusual to find that only one particular work relationship structure can be fitted to these tasks.

This element of choice and the mutual influence of technology and the social system may both be illustrated from the Tavistock studies, made over several years, of work organization in British deep seam coal mining. These indicated the main features of two very different forms of organization that have both been operated economically within the same seam and with identical technology.

The conventional system combines a complex formal structure with simple work roles: the composite system combines a simple formal structure with complex work roles. In the former the miner has a commitment to only a single part task and enters into only a very limited number of unvarying social relations that are sharply divided between those within his particular task group and those who are outside. With those 'outside' he shares no sense of belongingness and he recognizes no responsibility to them for the consequences of his actions. In the composite system the miner has a commitment to the whole group task and consequently finds himself drawn into a variety of tasks in co-operation with different members of the total group; he may be drawn into any task on the coal-face with any member of the total group.

That two such contrasting social systems can effectively operate the same technology is clear enough evidence that there exists an element of choice in designing a work organization. However, it is not a matter of indifference which form of organization is selected. As has already been stated, the technological system sets certain requirements of its social system and the effectiveness of the total production system will depend upon the adequacy with which the social system is able to cope with these requirements. Although alternative social systems may survive in that they are both accepted as 'good enough' (Simon, 1957) this does not preclude the possibility that they may differ in effectiveness. In this case the composite systems consistently showed a superiority over the conventional in terms of production and costs.

This superiority reflects, in the first instance, the more adequate coping in the composite system with the task requirements. The constantly changing underground conditions require that the already complex sequence of mining tasks undergo frequent changes in the relative magnitudes and even the order of these tasks. These conditions optimally require the internal flexibility possessed in varying degrees by the composite systems. It is difficult to meet variable task requirements with any organization built on a rigid division of labour. The only justification for a rigid division of labour is a technology which demands specialized non-substitute skills and which is, moreover, sufficiently superior, as a technology, to offset the losses due to rigidity. The conventional longwall cutting system has no such technical superiority over the composite to offset its relative rigidity—its characteristic inability to cope with changing conditions other than by increasing the stress placed on its members, sacrificing smooth cycle progress or drawing heavily upon the negligible labour reserves of the pit.

The superiority of the composite system does not rest alone in more adequate coping with the tasks. It also makes better provision to the personal requirements of the miners. Mutually supportive relations between task groups are the exception in the conventional system and the rule in the composite. In consequence, the conventional miner more frequently finds himself without support from his fellows when the strain or size of his task requires it. Crises are more likely to set him against his fellows and hence worsen the situation.

Similarly, the distribution of rewards and statuses in the conventional system reflects the relative bargaining power of different roles and task groups as much as any true differences in skill and effort. Under these conditions of disparity between effort and reward any demands for increased effort are likely to create undue stress.

These findings were replicated by experimental studies in textile mills in the radically different setting of Ahmedabad, India (Rice, 1958).

However, two possible sources of misunderstanding need to be considered:

(1) our findings do not suggest that work group autonomy should be maximized in all productive settings. There is an optimum level of grouping which can be determined only by analysis of the requirements of the technological system. Neither does there appear to be any simple relation between level of mechanization and level of grouping. In one mining study we found that in moving from a hand-filling to a machine-filling technology, the appropriate organization shifted from an undifferentiated composite system to one based on a number of partially segregated task groups with more stable differences in internal statuses;

(2) nor does it appear that the basic psychological needs being met by grouping are workers' needs for friendship on the job, as is frequently postulated by advocates of better 'human relations' in industry. Grouping produces its main psychological effects when it leads to a system of work roles such that the workers are primarily related to each other by way of the requirements of task performance and task inter-dependence. When this task orientation is established the worker should find that he has an adequate range of mutually supportive roles (mutually supportive with respect to performance and to carrying stress that arises from the task). As the role system becomes more mature and integrated, it becomes easier for a worker to understand and appreciate his relation to the group. Thus in the comparison of different composite mining groups it was found that the differences in productivity and in coping with stress were not primarily related to differences in the level of friendship in the groups. The critical prerequisites for a composite system are an adequate supply of the required special skills among members of the group and conditions for developing an appropriate system of roles. Where these prerequisites have not been fully met, the composite system has broken down or established itself at a less than optimum level. The development of friendship and particularly of mutual respect occurs in the composite systems but the friendship tends to be limited by the requirements of the system; it did not assume unlimited disruptive forms such as were observed

in conventional systems and were reported by Adams (1953) to occur in certain types of bomber crews.

Boundary Keeping

The textile studies (Rice, op. cit.) yielded the additional finding that *supervisory roles* are best designed on the basis of the same type of socio-technical analysis. It is not enough simply to allocate to the supervisor a list of responsibilities for specific tasks and perhaps insist upon a particular style of handling men. The supervisory roles arise from the need to control and co-ordinate an incomplete system of men-task relations. Supervisory responsibility for the specific parts of such a system is not easily reconcilable with responsibility for overall aspects. The supervisor who continually intervenes to do some part of the productive work may be proving his willingness to work but is also likely to be neglecting his main task of controlling and co-ordinating the system so that the operators are able to get on with their jobs with the least possible disturbance.

Definition of a supervisory role presupposes analysis of the system's requirements for control and co-ordination and provision of conditions that will enable the supervisor readily to perceive what is needed of him and to take appropriate measures. As his control will in large measure rest on his control of the boundary conditions of the system—those activities relating to a larger system—it will be desirable to create 'unified commands' so that the boundary conditions will be correspondingly easy to detect and manage. If the unified commands correspond to natural task groupings, it will also be possible to maximize the autonomous responsibility of the work group for internal control and co-ordination, thus freeing the supervisor for his primary task.

The re-organization achieved by Rice was reflected in a significant and sustained improvement in mean percentage efficiency and a decrease in mean percentage damage.

The difference between the old and the new weaving shed organizations does not rest only in the relative simplicity of the latter (although this does reflect less confusion of responsibilities) but also in the emergence of such clearly distinct areas of command which contain within themselves a relatively independent set of work roles together with the skills necessary to govern their task boundaries. In like manner the induction and training of new members was recognized as a boundary condition for the entire shed and located directly under shed management instead of being scattered throughout subordinate commands. Whereas the former organization had been maintained in a steady state only by the constant and arduous efforts of management, the new one proved to be inherently stable and self-correcting, and consequently freed management to give more time to their primary taks and also to manage a third shift.

The Environment

Similarly, the primary task in managing the enterprise as a whole is to relate the total system to its environment and is not in internal regulation *per se.* This does not mean that managers will not be involved in internal problems but that such involvement will be oriented consciously or unconsciously to certain assumptions about the external relations of the enterprise.

This contrasts with the common postulate of the structural-functional theories that 'the basic need of all empirical systems is the maintenance of the integrity and continuity of the system itself' (Selznick, 1948). It contrasts also with an important implication of this postulate, namely, that the primary task of management is 'continuous attention to the possibilities of encroachment and to the forestalling of threatened aggressions or deleterious consequences from the actions of others'. In industry this represents the special and limiting case of a management that takes for granted a previously established definition of its primary task and assumes that all they have to do, or can do, is sit tight and defend their market position. This is, however, the common case in statutorily established bodies and it is on such bodies that recent studies of bureaucracy have been largely carried out.

In general the leadership of an enterprise must be willing to break down an old integrity or create profound discontinuity if such steps are required to take advantage of changes in technology and markets. The very survival of an enterprise may be threatened by its inability to face up to such demands, as for instance, switching the main effort from production of processed goods to marketing or from production of heavy industrial goods to consumer goods. Similarly, the leadership may need to pay 'continuous' attention to the possibilities of making their own encroachments rather than be obsessed with the possible encroachments of others.

Considering enterprises as 'open socio-technical systems' helps to provide a more realistic picture of how they are both influenced by and able to act back on their environment. It points in particular to the various ways in which enterprises are enabled by their structural and functional characteristics ('system constants') to cope with the 'lacks' and 'gluts' in their available environment. Unlike mechanical and other inanimate systems they possess the property of 'equi-finality'; they may achieve a steady state from differing initial conditions and in differing ways (von Bertalanffy, op. cit.). Thus in coping by internal changes they are not limited to simple quantitative change and increased uniformity but may, and usually do, elaborate new structures and take on new functions. The cumulative effect of coping mainly by *internal* elaboration and differentiation is generally to make the system independent of an increasing range of the predictable fluctuations in its supplies and outlets. At the same time, however, this process ties down in specific ways more and more of its capital, skill and energies and renders it less able to cope with newly emergent and unpredicted changes that challenge the primary ends of the enterprise. This process has been traced out in a great many empirical studies of bureaucracies (Blau, op. cit., Merton, op. cit., Selznick, 1949).

However, there are available to an enterprise other aggressive strategies that seek to achieve a steady state by transforming the environment. Thus an enterprise has some possibilities for moving into new markets or inducing changes in the old; for choosing differently from amongst the range of personnel, resources and technologies offered by its environment or training and making new ones; and for developing new consumer needs or stimulating old ones.

Thus, arising from the nature of the enterprise as an open system, management is concerned with 'managing' both an internal system and an external environment. To regard an enterprise as a closed system and concentrate upon management of the 'internal enterprise' would be to expose the enterprise to the full impact of the vagaries of the environment.

If management is to control internal growth and development it must in the first instance control the 'boundary conditions'—the forms of exchange between the enterprise and its environment. As we have seen most enterprises are confronted with a multitude of actual and possible exchanges. If resources are not to be dissipated the management must select from the alternatives a course of action. The casual texture of competitive environments is such that it is extremely difficult to survive on a simple strategy of selecting the best from among the alternatives immediately offering. Some that offer immediate gain lead nowhere, others lead to greater loss; some alternatives that offer loss are avoidable, others are unavoidable if long run gains are to be made. The relative size of the immediate loss or gain is no sure guide as to what follows. Since also the actions of an enterprise can improve the alternatives that are presented to it, the optimum course is more likely to rest in selecting a strategic objective to be achieved in the long run. The strategic objective should be to place the enterprise in a position in its environment where it has some assured conditions for growth—unlike war the best position is not necessarily that of unchallenged monopoly. Achieving this position would be the *primary task* or overriding mission of the enterprise.

In selecting the primary task of an enterprise, it needs to be borne in mind that the relations with the environment may vary with (a) the productive efforts of the enterprise in meeting environmental requirements; (b) changes in the environment that may be induced by the enterprise and (c) changes independently taking place in the environment. These will be of differing importance for different enterprises and for the same enterprises at different times. Managerial control will usually be greatest if the primary task can be based on productive activity. If this is not possible, as in commerce, the primary task will give more control if it is based on marketing than simply on foreknowledge of the independent environmental changes. Managerial control will be further enhanced if the primary task, at whatever level it is selected, is such as to enable the enterprise to achieve *vis-á-vis* its competitors, a *distinctive competence.* Conversely, in our experience, an enterprise which has long occupied a favoured position because of distinctive productive competence may have grave difficulty in recognizing when it is losing control owing to environmental changes beyond its control.

As Selznick (1957) has pointed out, an appropriately defined primary task offers stability and direction to an enterprise, protecting it from adventurism or costly drifting. These advantages, however, as he illustrates, may be no more than potential unless the top management group of the organization achieves solidarity about the new primary task. If the vision of the task is locked up in a single man or is the subject of dissension in top management it will be subject to great risk of distortion and susceptible to violent fluctuations. Similarly, the enterprise as a whole needs to be re-oriented and reintegrated about this primary task. Thus, if the primary task shifts from heavy industrial goods to durable consumer goods it would be necessary to ensure that there is a corresponding shift in values that are embodied in such sections as the sales force and design department.

References

ABRAMSON, J. A. and SCOTT, J. (1968). *The power of an image to affect commitment to social action.* Saskatoon: Canadian Centre for Community Studies.

ACKOFF, R. L. (1949). On a science of ethics. *Philos. & phenomenological research,* **9**, 663-672.

ACKOFF, R. L. (1959). Games, decisions and organizations. In *General systems yearbook.*

ACKOFF, R. L. (1966). The meaning of strategic planning. *McKinsey qtly.,* **3**, 48-62.

ACKOFF, R. L. (1968). Operational research and national science policy. In de Reuck, A. (ed.) *Decision making in national science policy.* London: Churchill.

ACKOFF, R. L. (1969a). Institutional functions and societal needs. In Jantsch, E. (ed.). *Perspectives in planning.* Paris: O.E.C.D.

ACKOFF, R. L. (1969b). *A concept of corporate planning.* New York and London: Wiley.

ACKOFF, R. L. (1970). *A concept of corporate planning.* New York: Wiley–Inter Science.

ACKOFF, R. L. and EMERY, F. E. (to be published, 1972). *On purposeful systems.* Ill.: Aldine Press.

ADAMS, S. (1953). Status congruency as a variable in small group performance. *Social forces,* **32**, 16-22.

ADLER, A. G. (1958). Some ideas toward a sociology of the concentration camps. *Social forces.* **68**, 513-523.

ALMOND, G. and COLEMAN, J. (eds.). (1960). *The politics of the developing areas.* Princeton: Princeton University Press.

ANGYAL, A. (1941). *Foundations for a science of personality.* Cambridge, Mass.: Harvard University Press. (Re-issued 1958).

ANGYAL, A. (1966). *Neurosis and treatment.* New York and London: Wiley.

ARENDT, H. (1958). *The human condition.* Chicago: University of Chicago Press; Cambridge: Cambridge University Press.

ARGYRIS, C. (1964). *Integrating the individual and the organization.* New York and London: Wiley.

ARNHEIM, R. (1954). *Art and visual perception.* Berkeley: University of California Press.

ASCH, S. E. (1956). Studies of indepedence and conformity. *Psychological monographs,* **70,** No. 9.

ASHBY, W. R. (1956). *Introduction to cybernetics.* London: Chapman & Hall.

ASHBY, W. R. (1960). *Design for a brain.* Second edition. London: Chapman & Hall.

AXELROD, J. *et al.* (1969). *Search for relevance.* San Francisco: Jossey-Bass.

AYRES, R. U. (1966). On technological forecasting. In *Working papers of the commission on the year 2000.* Am. Acad. Arts & Sci.

BARKER, R. G. *et al.* (1946). *Adjustment to physical handicap and illness.* New York: SSRC Bulletin 55.

BARNARD, C. I. (1948). *The functions of the executive.* Cambridge, Mass.: Harvard University Press.

BARTH, F. (1965). *Three modes of anthropology.* London: Royal Society.

BELL, D. (1965). Twelve modes of prediction. In Gould, J. (ed.). *Penguin survey of the social sciences.* London: Penguin Books.

BELL, D. (1967). The year 2000—the trajectory of an idea. *J. am. acad. arts & sci.* **96,** 639-651.

BENNIS, W. G. and SLATER, P. E. (1968). *The temporary society.* New York: Harper & Row.

BERLE, A. A. and MEANS, G. C. (1932). *The modern corporation and private property.* New York: Macmillan.

BION, W. R. (1947-51). Various papers in *Hum. relat.* collected in *Experiences in groups, and other papers (1961).* London: Tavistock Publications; New York: Basic Books.

BION, W. R. and RICKMAN, J. (1943). Intra-group tensions in therapy. *Lancet,* 2, 678-681.

BLAU, P. (1955). *The dynamics of bureaucracy: a study of interpersonal relations in two government agencies.* Chicago: University of Chicago Press.

BOTT, E. (1957). *Family and social network.* London: Tavistock Publications.

BOULDING, K. (1956). *The image: knowledge in life and society.* Ann Arbor: University of Michigan Press.

BOULDING, K. (1966). Conflict management as a learning process. In de Reuck, A. (ed.). *Conflict in society.* London: Churchill.

BOWLBY, J. (1969). *Attachment and loss, Vol. 1: attachment.* London: Tavistock Publications; New York: Basic Books.

BRIDGER, H. (1946). The Northfield experiment. *Bull. menninger clin.* **10,** 71-76.

BRONOWSKI, J. (1961). *Science and human values.* London: Hutchinson.

BRUNER, J. (1964). Presidential address. *J. soc. issues,* **20,** 21-33.

BURNS, T. and STALKER, G. (1961). *The management of innovation.* London: Tavistock Publications.

CANADIAN WELFARE COUNCIL (1967). *A comprehensive statement on social welfare for Canada.* Ottawa: Canadian Welfare Council.

CANTRIL, H. (1941). *The psychology of social movements.* New York: Wiley.

CANTRIL, H. (1965). *The patterns of human concern.* New Jersey: Rutgers University Press.

CHANCE, M. R. A. (1960). Köhler's chimpanzees–how did they perform? *Man,* **60,** 130-135.

CHEIN, I. (1944). The logic of prediction: some observations on Dr. Sarbim's exposition. *Psychol. rev.* **52,** 175-179.

CHEIN, I. (1947). Towards a science of morality. *J. soc. psychol.* **25,** 235-238.

CHEIN, I. (1948). The genetic factor in a historical psychology. *J. gen. psychol.* **26,** 151-172.

CHEIN, I. (1954). The environment as a determinant of behavior. *J. social psychol.* **39,** 115-127.

CHEVALIER, M. (1966). *A wider range of perspectives in the bureaucratic structure.* Working paper. Ottawa: Commission on Bi-Lingualism and Bi-Culturalism.

CHEVALIER, M. (1967). *Stimulation of needed social science research for Canadian water resource problems.* Ottawa: Privy Council Science Secretariat.

CHEVALIER, M. (1968). *Interest group planning.* Ph.D. thesis, University of Pennsylvania.

CHURCHMAN, C. W. (1961). *Prediction and optimal decision.* New York: Prentice Hall.

CHURCHMAN, C. W. (1966). *On the ethics of large-scale systems.* Berkeley: Space Sciences Laboratory, University of California.

CHURCHMAN, C. W. and ACKOFF, R. L. (1949). The democratization of philosophy. *Science & Society,* **13**, 327-339.

CHURCHMAN, C. W. *et al.* (1967). *Experiments on inquiring systems.* Berkeley: Social Sciences Group, Space Sciences Laboratory, University of California.

CHURCHMAN, C. W. and EMERY, F. E. (1966). On various approaches to the study of organizations. In Lawrence, J. R. (ed.). *Operational research and the social sciences.* London: Tavistock Publications.

CHURCHMAN, C. W. and VERHULST, M. (eds.). (1960). *Management sciences: models and techniques.* London: Pergamon Press.

CLEGG, H. (1960). *Industrial democracy.* Oxford: Blackwell.

COMMITTEE ON THE NEXT THIRTY YEARS, BRITISH SOCIAL SCIENCE RESEARCH COUNCIL. (1968). *Forecasting and the social sciences.* London: Heinemann.

COHN, N. (1957). *The pursuit of the millenia.* London: Secker & Warburg.

COWAN, T. A. (1965). *Non-rationality in decision theory.* Berkeley: Space Sciences Laboratory, University of California.

CRICHTON, C. (ed.). (1966). *Interdependence and uncertainty.* London: Tavistock Publications.

CROZIER, M. (1964). *The bureaucratic phenomenon.* London: Tavistock Publications; Chicago: University of Chicago Press.

CROZIER, M. (1966). Attitudes and beliefs in national planning. In Gross, B. M. (ed.). *Action under planning.* New York: McGraw-Hill.

DE BIE, P. (1970). Problem-focussed research. In *Main trends of research in the social and human sciences.* Paris: Mouton/Unesco.

DE GROOT, A. (1965). *Thought and choice in chess.* The Hague: Mouton.

DELEGATION GENERALE A LA RECHERCHE SCIENTIFIQUE ET TECHNIQUE (DGRST). (1965). Rapport du groupe 'sciences humaines'. Préparation du Véme plan. Paris: DGRST.

DEUTSCH, H. (1942). Some forms of emotional disturbances and their relationship to schizophrenia. *Psychoanal. qtly.* 2, 301.

DEUTSCH, M. (1954). Field theory in social psychology. In Lindzey, G. (ed.). *Handbook of social psychology,* Vol. 1, Ch. 5. Cambridge, Mass.: Addison Wesley.

DICKS, H. V. (1950). Personality traits and national socialist ideology. *Hum. relat.* **3**, 111-154.

DICKS, H. V. (1952). Observations on contemporary Russian behaviour. *Hum. relat.* **5**, 111-175.

DREYFUS, H. L. (1965). *Alchemy and artificial intelligence.* Santa Monica, Calif.: Rand Corporation.

DRUCKER, P. F. (1952). The employee society. *Am. sociol. rev.* **58**, 358-363.

DRUCKER, P. F. (1957). *Landmarks of tomorrow.* New York: Harper.

DUHL, L. D. (ed.). (1963). *The urban condition.* New York: Basic Books.

EMERY, F. E. and TRIST, E. L. (1965). The causal texture of organizational environments. *Hum. relat.* **18**, 21-32.

EMERY, F. E. (1967). The democratization of the work-place. *Manpower and applied psychol.* **1**, 118-130.

EMERY, F. E. (1967). The next thirty years: concepts, methods and anticipations. *Hum. relat.* **20**, 199-237.

EMERY, F. E., HILGENDORF, E. L. and IRVING, B. L. (1968). *The psychological dynamics of smoking.* London: Tobacco Research Council.

EMERY, F. E. and THORSRUD, E. (1969). *Form and content in industrial democracy.* London: Tavistock Publications. (Norwegian edition, 1964, Oslo University Press).

EMERY, F. E. (ed.). (1969). *Systems thinking.* London: Penguin.

EMERY, F. E. and WILSON, I. H. (to appear 1971). Mao's model of China's future. *Current scene.*

ERICKSON, J. (1970). *Defence and technology: some questions of the management of weapons technology.* Paper given at conference on the impact of science and technology, Edinburgh University. (Proceedings forthcoming).

ERIKSON, E. H. (1968). *Identity: youth and crisis.* London: Faber & Faber; New York: Norton.

EVAN, W. M. (1966). The organizational set: toward a theory of inter-organizational relations. In Thompson, J. D. (ed.). *Approaches to organizational design.* Pittsburgh: University of Pittsburgh Press.

FAIRBAIRN, W. R. D. (1952). *Psychoanalytic studies of the personality.* London: Tavistock Publications.

FERGUSSON, C. E. and PFOUTTS, R. W. (1962). Learning and expectations in dynamic duopoly behavior. *Behav. sci.* 7, 223-237.

FRANCK, L. K. (1967). The need for a new political theory. *J. am. acad. arts & sci.* **96**, 809-816.

FRANKFORT, H. *et al.* (1949). *Before philosophy.* London: Pelican.

FRIEDMANN, J. and MILLER, J. (1965). The urban field. *J. amer. inst. planners,* **31**, 312-320.

FRIEND, J. K. and JESSOP, W. N. (1969). *Local government and strategic choice. An operational research approach to the process of public planning.* London: Tavistock Publications.

FROMM, E. (1950). *Escape from freedom.* London: Routledge & Kegan Paul.

GALBRAITH, J. K. (1967). *The new industrial state.* New York: Houghton Mifflin.

GEDDES, SIR PATRICK, (1968). *Cities in Evolution.* (New edition). London: Ernest Benn.

GIBSON, Q. (1960). *The logic of social inquiry.* London: Routledge & Kegan Paul.

GOLDSCHMIDT, W. (1959). *Understanding human society.* London: Routledge & Kegan Paul.

GOULDNER, A. W. (1954). *Patterns of industrial bureaucracy.* Glencoe, Ill.: Free Press; London: Routledge & Kegan Paul (1955).

GROSS, B. M. (1966). The state of the nation. In Bauer, R. (ed.). *Social indicators.* Cambridge, Mass.: M.I.T. Press; London: Tavistock Publications.

GROSS, B. M. (ed.). (1966). *Action under planning.* New York: McGraw-Hill.

GROSS, B. M. (1967). The city of man: a social systems reckoning. In Ewald, W. R. (ed.). *Environment for man.* Bloomington: Indiana University Press.

GROSS, B. M. (1968a). *An overview of change in America.* Unpublished report. New York: Twentieth Century Fund.

GROSS, B. M. (ed.). (1968b). *A great society?* New York: Basic Books.

HAGEN, E. E. (1960). *On the theory of social change: how economic growth begins.* London: Tavistock Publications; Homewood, Ill.: Dorsey Press (1962).

HAMILTON, G. V. (1911). A study of trial and error reactions in mammals. *J. animal behavior,* Vol. 1. pp. 33-66 and in Riopelle, A. J. (1967). *Animal problem solving,* London: Penguin.

HARRINGTON, M. (1965). *The accidental century.* New York: Macmillan; London: Penguin Books (1968).

HEGEL, C. W. F. (1949). *Phenomenology of the mind.* London: Allen & Unwin.

HEIDER, F. (1958). *The psychology of interpersonal relations.* New York: Wiley.

HERBST, P. G. (1954). The analysis of social flow systems. *Hum. relat.* 7, 327-336.

HIGGIN, G. W., EMERY, F. E. and TRIST, E. L. (1959). *Communications in the National Farmers' Union of England and Wales.* London: Tavistock Institute.

HIGGIN, G. W. and JESSOP, N. (1963). *Communications in the building industry.* London: Tavistock Publications.

HIRSCHMAN, A. O. and LINDBLOM, C. E. (1962). Economic development, research and development, policy making: some convergent views. *Behavl. sci.* 7, 211-222.

HOLFORD, Lord (1965). *The build environment.* London: Tavistock Publications.

HOMANS, G. C. (1958). Social behavior as exchange. *A. J. sociol.* 63, 597-606.

INSTITUTE FOR OPERATIONAL RESEARCH (IOR) (1967). *The first four years.* London: Tavistock Institute.

JACOBY, N. H. (1967). *Canada's tax structure and economic goals.* York, Ont.: York University, Faculty of Administrative Studies.

JAQUES, E. (1951). *The changing culture of a factory.* London: Tavistock Publications; New York: Dryden (1952).

JAQUES, E. (1956). *Measurement of responsibility; a study of work, payment, and individual capacity.* London: Tavistock Publications; Cambridge, Mass.: Harvard University Press.

JAQUES, E. (1960). *Equitable payment; a general theory of work, differential payment and individual progress.* London: Heinemann; New York: Wiley (1961).

JAQUES, E. (1965). Death and the mid-life crisis. *I. j. psychoanal.* 46, 502-514.

JORDAN, N. (1968). *Themes in speculative psychology.* London: Tavistock Publications.

JUNG, C. G. (1953). Two essays on analytical psychology. In *Collected works vol. 7.* London: Routledge & Kegan Paul; New York: Pantheon Books.

KAHN, H. and WIENER, A. J. (1967). *The year 2000.* New York: Macmillan.

KAPLAN, A. (1964). *The conduct of inquiry.* San Francisco: Chandler.

KENDRICK, J. W. (1967). *Studies in the national income accounts. 47th annual report of the National Bureau of Economic Research.* New York: NBER.

KLEIN, M. (1948). *Contributions to psycho-analysis 1921-1945.* London: Hogarth.

KLEIN, M. (1957). *Envy and gratitude.* London: Tavistock Publications.

KLEIN, M. and RIVIERE, J. (1937). *Love, hate and reparation.* London: Hogarth.

KÖHLER, W. (1927). *Mentality of apes.* London: Methuen.

KUHN, T. S. (1962). *The structure of scientific revolutions.* Chicago: University of Chicago Press.

LAING, R. D. (1960). *The divided self; a study of sanity and madness.* London: Tavistock Publications; Chicago: Quadrangle Books.

LAING, R. D. (1961). *The self and others; further studies in sanity and madness.* London: Tavistock Publications; Chicago: Quadrangle Books (1962).

LAING, R. D. (1967). *The politics of experience.* London: Penguin Books.

LAING, R. D. and ESTERSON, A. (1964). *Sanity, madness and the family. Vol. 1 Psychiatry—cases, clinical reports, statistics.* London: Tavistock Publications; New York: Basic Books (1965).

LAWRENCE, J. R. (ed.). (1966). *Operational research and the social sciences.* London: Tavistock Publications.

LAWRENCE, P. R. and LORSCH, J. W. (1967). *Organization and Environment.* Boston: Division of Research, Graduate School of Business Administration, Harvard University.

LEVI-STRAUSS, C. (1960). *The scope of anthropology.* English translation London: Cape 1967.

LEWIN, K. (1936a). *A dynamic theory of personality.* New York: McGraw-Hill.

LEWIN, K. (1936b). *Topological psychology.* New York: McGraw-Hill.

LEWIN, K. (1951). *Field theory in social science.* (Cartwright, D. (ed.).) New York: Harper; London: Tavistock Publications.

LIDDELL-HART, B. H. (1944). *Thoughts on war.* London: Faber.

LIKERT, R. (1961). *New patterns of management.* New York: McGraw-Hill.

LIKERT, R. (1967). *The human organization: its management and value.* New York: McGraw-Hill.

McCLELLAND, D. C. (1961). *The achieving society.* Princeton: D. van Nostrand.

McGRANAHAN, D. V. (1946). A comparison of social attitudes among American and German youth. *J. abnorm. soc. psychol.* **41**, 245-257.

McGREGOR, D. (1960). *The human side of enterprise.* New York: McGraw-Hill.

McLUHAN, M. (1964). *Understanding media.* New York: McGraw-Hill; London: Routledge & Kegan Paul.

McWHINNEY, W. (1968). Organizational form, decision modalities and the environment. *Hum. relat.* **21**, 269-281.

MAIN, T. F. (1946). The hospital as a therapeutic institution. *Bull. menninger clin.* **10**, 66-70.

MARCUSE, H. (1956). *Eros and civilization.* London: Routledge & Kegan Paul.

MARCUSE, H. (1964). *One dimensional man.* London: Routledge & Kegan Paul.

MASLOW, A. H. (1954). *Motivation and personality.* New York: Harper.

MASLOW, A. H. (1967). *Unpublished manuscript of the B values.*

MERTON, R. K. (1949). *Social theory and social structure.* First edition. Glencoe, Ill.: Free Press.

MUMFORD, L. (1961). *The city in history.* New York: Harcourt Brace & World.

MYRDAL, G. (1941). *The american dilemma,* Vol I & II. New York: Harper & Row.

NEUMANN, E. (1966). *Art and the creative unconscious.* New York: Harper.

NEWCOMB, T. M. (1947). Autistic hostility and social reality. *Hum. relat.* **1**, 43-69.

PARETO, V. (1935). *The mind and society.* New York: Harcourt, Brace.

PAVLOV, I. P. (1928). *Lectures on conditioned reflexes.* New York: International Publishers.

PEPPER, S. C. (1961). *World hypotheses.* Berkeley: University of California Press.

PERLMUTTER, H. V. (1965). L'entreprise internationale trois conceptions. *Revue économique et sociale,* **2**, 1-14.

PERLMUTTER, H. V. (1969). Some management problems in spaceship earth: the mega firm and the global industrial estate. In Scott, W. P. and le Breton, P. P. (eds.). *Managing complex organizations.* Seattle: University of Washington Press.

POLANYI, M. (1967). The growth of science in society. *Minerva,* **5**, 533-545.

PRICE, D. DE S. (1961). *Science since Babylon.* New Haven: Yale University Press.

PRICE, D. K. (1965). *The scientific estate.* Cambridge, Mass.: Belknap.

RAPOPORT, R. and RAPOPORT, R. N. (1969). The dual career family. *Hum. relat.* **22**, 3-30.

REES, J. R. (1945). *The shaping of psychiatry by war.* New York: Norton.

RICE, A. K. (1958). *Productivity and social organization. The Ahmedabad experiment.* London: Tavistock Publications.

RICE, A. K. and TRIST, E. L. (1952). Institutional and sub-institutional determinants of change in labour turnover (the Glacier project–VIII). *Hum. relat.* **5**, 347-372.

RIESMAN, D. (1950). *The lonely crowd.* New Haven: Yale University Press.

RIESMAN, D. (1958). Leisure and work in post-industrial society. In Larrabee, E. and Mayershon, R. (eds.). *Mass leisure.* Glencoe, Ill.: Free Press.

RUTHERFORD, A. (1969). *Towards a bargaining model.* Unpublished manuscript, Yale Law School.

SARTRE, J. P. (1964). Questions of method. In Laing, R. D. and Cooper, D. G. (eds.). *Reason and violence: a decade of Satre's philosophy 1950-1960.* London: Tavistock Publications.

SCHON, D. (1963). *The displacement of concepts.* London: Tavistock Publications.

SCHULTZ, T. (1961). Investment in human capital. *Am. econ. r.* **51**, 1-17.

SCHUTZENBERGER, M. P. (1954). A tentative classification of goal-seeking behaviours. *J. ment. sci.* **100**, 97-102.

SELZNICK, P. (1948). Foundations of the theory of organization. *Amer. soc. rev.* **13**, 25-35.

SELZNICK, P. (1949). *TVA and the grass roots.* Berkeley: University of California Press.

SELZNICK, P. (1957). *Leadership in administration.* Evanston, Ill.: Row Peterson.

SHELL REFINING CO. LTD. (1966). *Statement of company objectives and management philosophy.* London: Shell Centre.

SHONFIELD, A. (1966). *Modern capitalism.* London: Oxford University Press.

SIMON, H. A. (1956). Rational choice and the structure of the environment. *Psychol. rev.* **63**, 129-138.

SIMON, H. A. (1957). *Models of man.* New York: Wiley.

SNOW, Lord. (1961). *Science and government.* Cambridge: Cambridge University Press.

SOMMERHOFF, G. (1950). *Analytical biology.* London: Oxford University Press.

SOMMERHOFF, G. (1969). The abstract characteristics of living systems. In Emery, F. E. (ed.). *Systems thinking.* Harmondsworth, Middlesex: Penguin Books.

SPRANGER, E. (1928). *Types of men: the psychology and ethics of personality.* Fifth German edition, trans. Pigors. Halle: Niemeyer.

STONE, P. J. *et al.* (1966). *The general inquirer.* Cambridge, Mass.: M.I.T. Press.

STRAUSS, A. *et al.* (1964). The hospital and its negotiated order. In Friedson, E. (ed.). *The hospital in modern society.* New York: The Free Press.

TAVISTOCK INSTITUTE OF HUMAN RELATIONS (1964). *Social research and a national policy for science.* London: Tavistock Publications.

THORNDIKE, E. L. (1911). *Animal intelligence.* New York: Hafner.

TODA, M. (1962). The design of a fungus-eater: a model of human behaviour in an unsophisticated environment. *Behav. sci.* **7**, 164-183.

TOLMAN, E. G. and BRUNSWICK, E. (1935). The organism and the causal texture of the environment. *Psychol. rev.* **42**, 43-77.

TOMKINS, S. S. (1962/3). *Imagery, affect, consciousness.* Vols. I and II. New York: Springer.

TRIST, E. L. (1970). Science policy and development of research in social science–the organization and financing of research. Section 3 in *Main trends in the social and human sciences.* Paris: Mouton/Unesco. pp. 693-811.

TRIST, E. L. and BAMFORTH, K. W. (1951). Some social and psychological consequences of the longwall method of coal-getting. *Hum. relat.* **4**, 3-38.

TRIST, E. L., HIGGIN, G. W., MURRAY, H. and POLLOCK, A. B. (1963). *Organizational choice: capabilities of groups at the coal face under changing technologies.* London: Tavistock Publications.

TROTTER, W. (1916). *Instincts of the herd in peace and war.* London: Fisher Unwin.

UNESCO (1965). *Science policy and organization of scientific research in the Czechoslovak Socialist Republic.* Science Policy Studies and Documents No. 2. Paris: Unesco.

UNESCO (1968). *National science policies of the U.S.A.* Science Policy Studies and Documents No. 10. Paris: Unesco.

U.N. DEPARTMENT OF ECONOMIC AND SOCIAL AFFAIRS (1963). *1963 Report on the world situation.* New York: United Nations.

VICKERS, Sir GEOFFREY (1959). *The undirected society.* Toronto: University of Toronto Press.

VICKERS, Sir GEOFFREY (1965). *The art of judgment.* London: Chapman & Hall. New York: Basic Books (1966).

VICKERS, Sir GEOFFREY (1968). *Value systems and social process.* London: Tavistock Publications; New York: Basic Books.

VICKERS, Sir GEOFFREY (1970). *Freedom in a rocking boat.* London: Allen Lane the Penguin Press; New York: Basic Books.

Von BERTALANFFY, L. (1950). The theory of open systems in physics and biology. *Science,* **3**, 23-29.

WALTON, R. E. and McKERSIE, R. B. (1965). *A behavioral theory of labor negotiations: an analysis of a social interaction system.* New York: McGraw-Hill.

WARNER, W. L. and LOW, J. O. (1947). *The social system of the modern factory.* New Haven: Yale University Press.

WAY, M. (1967). The road to 1977. *Fortune,* January 1967.

WEBER, M. (1947). *The protestant ethic and the spirit of capitalism.* London: Oxford University Press.

WERTHEIMER, M. (1959). *Productive thinking.* Revised edition. New York: Harper.

WICKENDEN, E. (1965). *Social welfare in a changing world.* Washington: U.S. Public Health Service.

WILSON, A. T. M., TRIST, E. L. and CURLE, A. (1952). Transitional communities and social reconnection. A study of civil resettlement of British prisoners of war. In Swanson, G. E. *et al.* (eds.). *Readings in social psychology.* Second edition. New York: Holt.

WINNICOTT, D. W. (1958). *Collected papers.* London: Tavistock Publications.

YOUNG, M. (1958). *The meritocracy.* London: Routledge & Kegan Paul.

ZENER, K. (1937). The significance of behavior accompanying conditioned salivary secretion for theories of the conditioned response. *Amer. J. of psychol.* **50**, 384.

Index

A

Adaption, 6
Analysis,
  based on the leading part, 20
  genotypical, 40
  linguistic usage of, 32, 33, 34, 37
  symbol of 24, 32, 37
  value in, 24, 32, 33, 37

C

Capacity for joy, 178
Common ground,
  function of, 188, 189
Competing systems,
  emerging social systems, 27
  extent of common parts, 28
  intrusion in, 27
  overlapping situation in, 28
  phases in, 26
  relative consonance of, 28
  state of, 26
  symptoms of, 27
  valence of, 28
  variance of, 27
Competitive relations, 182
Complexity reduction,
  difficulties in, 18
Concealment–Parasitism,
  energy requirements of, 25
  extrapolation prediction of, 24
  social processes in, 25
Co-presence, 49

D

Decision theory, 17
Directive correlation, 6, 19
Dissociation, 57, 65
Disturbed reactive environment, 48
Dynamic environments, 52
Dynamic field forces, 53

E

Ecological strategies, 185, 186
Environment, 220

F

Family cycle, 192
  income trajectories of, 195
Family status, 193
  nature of, 193
  trends in, 193

G

Genotypical analysis, 40
  (*See* analysis)
Growth process,
  curves of, 30
  Sigmoid character of, 29

H

Human relations, 105

I

Inter-dependence, 172, 177

L

Life careers, 115
Life cycle, 192

Life space, 191
changing role of, 191
nature of, 191
Life styles,
burden of choice, 201
changes in, 198
evolution of, 199
extension of, 199
financial needs of, 200

M

Methodology, 21, 26, 27, 31, 37

N

New Industrial State, 183

O

Organizational ecology, 113, 124
problems of, 114
theory of, 124
Organizational philosophies, 182

P

Personal impact, 190
Personal income,
reductions in, 197
reinforcements in, 195
trends in, 193, 194
Placid clustered environment, 45
assignment of step functions, 47
domain selection of, 46
hierarchy of strategies, 47
optimal location of, 46
strategic object of, 47
Placid random environment, 41
instrumentality of, 43
learning in, 43
planning of, 43
random distribution in, 44
storage capacity of, 43
Planning,
as process, 203
political process of, 205
social process of, 204
Pluralism, 206
Post-Industrial society, 120, 156
basis of, 158
cultural absence of, 172
drift to, 157
structural presence of, 156
Post-Industrial society—changes,
in economy, 161
in education, 164
in environmental context, 169
in family, 167
in leisure, 165
in occupation, 164
in power base, 160
in structure of, 161, 162
in unemployment, 165
Poverty programme, 110
Productive system, 39
Professional relationship, 110
field determination in, 112
function of, 110
fundamental data of, 111

R

Research and Development, 91
British Research Councils review of, 96
character of, 92
comparison of domain selection, 99
domain based on, 91
fundamental projects of, 99
main problems of, 99
mapping of, 94, 95
partial problems of, 99
problem councils in, 98
problems of, 92

S

Science,
appreciation of, 86
negative policy of, 84
positive policy of, 84
re-evaluating the role of, 83
social aspects of, 84
social problems of, 89
systems of, 88
Segmentation, 57, 63

Self-Actualization, 172, 175
Self-Expression, 172, 176
Social amplification, 103
Social design, 71
  basic principles of, 71
  choice of, 71
  control mechanisms of, 72
  environmental complexity of, 73
  redundancy of functions of, 72
Social development and innovation, 109
Social fields, 19
  environmental levels in, 38
  general characteristics of, 38
Social Science,
  active role in, 4
  as extension of choice, 8
  capabilities of, 4
  change processes in, 118
  conditions for, 116
  diffusion of, 116
  knowledge of, 116
  passive role of, 6
  planning in, 5
  social services in, 10
Socio-technical system, 211
  boundary conditions of, 214, 219
  closed system of, 211
  open system of, 211
  technical component of, 214, 215
Strategy,
  economic organization of, 38
  military organization of, 38
  overall characterization of, 38
Superficiality, 57, 59
Symbolic systems,
  changes in, 31
  state of, 31
System,
  environment relations in, 19
  environmental transformation of, 74
  identification of, 16
  management of, 73
  organizational matrix of, 76
  social accounting of, 17
System—*cont.*
  social physics of, 17
  strategic planning of, 77
  summary of, 77
System integration,
  depth of, 58
  dimensions of, 58
  means-end of, 58
  transverse of, 58

T

Temporal Gestalten, 13
  emergent overlap in, 15
  familiarity of, 13
  inclusiveness of, 14
  overlapping in, 12
  phase distance of, 13
Three-dimensional cultural model, 172
Turbulent environment, 52
  adaption to, 37
  adaptive response to, 68
  values of, 69
Turbulent fields, 52

V

Values,
  as indicators, 20
  system tendencies of, 20

W

Welfare and Development,
  adaptive planning in, 124
  analysis of, 131
  aspects of, 132
  basic characteristics of, 125
  forms of, 145
  function of, 129
  historical and contemporary aspects in, 137
  problems of, 123
  relation of, 120
  social ideas in, 132
  social practices of, 132
  social security in, 147